# HF SSB DX Basics

Steve Telenius-Lowe, PJ4DX

**Radio Society of Great Britain**

Published by the Radio Society of Great Britain, 3 Abbey Court, Fraser Road, Priory Business Park, Bedford MK44 3WH. Tel: 01234 832700. Web: www.rsgb.org

Published 2015.

© Radio Society of Great Britain, 2015. All rights reserved. No part of this publication may be reproduced, stored in a retrieval system, or transmitted, in any form or by any means, electronic, mechanical, photocopying, recording or otherwise, without the prior written permission of the Radio Society of Great Britain.

ISBN: 9781 9101 9315 0

Design and layout: Steve Telenius-Lowe, PJ4DX
Cover design: Kevin Williams, M6CYB
Production: Mark Allgar, M1MPA

Printed in Great Britain by Hobbs the Printer of Totton, Hampshire

*Publisher's Note:*
The opinions expressed in this book are those of the authors and are not necessarily those of the Radio Society of Great Britain. Whilst the information presented is believed to be correct, the publishers and their agents cannot accept responsibility for consequences arising from any inaccuracies or omissions.

## Acknowledgements

Some material in this book is adapted and re-edited from passages in *The Amateur Radio Operating Manual* (8th edition) edited by Mike Dennison, G3XDV, and Steve Telenius-Lowe, PJ4DX, and published by the RSGB in 2015. The original authors of that material were Don Field, G3XTT; Roger Balister, G3KMA, and Steve Telenius-Lowe, 9M6DXX (now PJ4DX). All material is copyright RSGB.

# Contents

| | Preface | 4 |
|---|---|---|
| 1 | What is HF SSB? | 5 |
| 2 | What is DX? | 20 |
| 3 | Planning Your Antenna | 27 |
| 4 | The SSB DXer's Transceiver | 45 |
| 5 | Transmitting SSB Speech | 59 |
| 6 | Propagation & SSB DX on HF | 71 |
| 7 | Working HF DX on SSB | 79 |
| 8 | Being DX | 93 |
| | Index | 96 |

# Preface

DXING – THE ART AND SCIENCE of making radio contact with the furthest reaches of the planet – is a hugely satisfying and addictive aspect of our hobby. It is practised by hundreds of thousands of radio amateurs around the world and these days more and more are joining in the fun on single sideband (SSB), the most popular voice communications mode on HF. The purpose of this book is to encourage more people – whether newcomers or the more experienced – to discover the delights of DXing using SSB on the HF bands.

I use the term "newcomers" in the widest sense. In the UK it includes Foundation and Intermediate licensees but, less obviously maybe, even the recently-licensed Full licensee. In the USA the term would include Technicians and newly-licensed General class licensees. But in this context it also encompasses those licensed for many years who so far have operated either exclusively on the VHF / UHF bands or those who might have operated on HF but who have had little or no interest in working DX.

Why should that be? Why should someone 'jump through the hoops' of the licensing system, but then be content only to talk to their local amateur neighbours on VHF or 80m? For some, that is all they *want* to do, and that is fine. No amount of encouragement by me or anyone else is going to make someone want to work DX if they have no interest in doing so in the first place. But for others, and I suspect the majority, they would *like* to work DX but have a feeling that, with a typical station of 100 watts and wire antennas, they cannot compete with those running linear amplifiers and large beam antennas. This book sets out to correct that misconception.

So give HF SSB DXing a try and and allow the DX 'bug' to bite! It is my hope that, if the DX bug *does* bite, such amateurs will go on to improve their stations: once they do so they will find that the DX becomes easier to work. Once the DX bug has bitten it is easy to become 'infected' and this can lead to a lifetime's interest in this fascinating aspect of the hobby.

*Steve Telenius-Lowe, PJ4DX*
*November 2015*

# 1 What is HF SSB?

AS MENTIONED IN THE PREFACE opposite, the main purpose of this book is to encourage 'newcomers' to dip their toes into the wonderful world of DXing on the HF bands. While the definition of newcomers does indeed include those who have been licensed for many years but who have not tried HF DXing before, the majority *will* probably be the newly-licensed.

As a result of the decision made at the International Telecommunication Union World Radio Conference in 2003, almost all the world's regulatory authorities no longer set a Morse code test as a prerequisite to obtaining an licence with access to the HF bands – the short-wave bands below 30MHz.

Unless the individual has had a specialist occupation that requires knowledge of Morse code (and such occupations are becoming fewer and fewer) the chances are that the newly-licensed amateur will not know Morse. It is a fact that tens of thousands of amateurs around the world who have become licensed since 2003 have little or no knowledge of Morse code, and these amateurs are now active on the HF bands.

Furthermore, since there is no longer a Morse test as a 'barrier' to HF operation, more amateurs will be coming on the air all the time who also have no knowledge of Morse code. The use of Morse will not die out, as some had predicted after the 2003 ITU decision, simply because there is no doubting that it *is* still a useful mode of communications. However, I think it is fair to say that single sideband – SSB – will become more and more predominant as time goes by and as more and more amateurs who do not have any knowledge of Morse become licensed.

It is for that reason that this book is an introduction to HF DXing from the perspective of the *SSB operator*. I make no apology for this: if the newcomer does not know Morse code there is little point in discussing the finer points of CW (Morse code) operating techniques. Many operators *do* go on to learn Morse and become proficient CW operators and for them there is an excellent book which has been available for many years [1]. But there is little for the SSB DX operator and even less for the newcomer to that field, so it is my hope that this book helps to fill that gap.

The next few sections are intended for those new operators who might have had little knowledge of radio before gaining their licence. I hope it may also be useful as a 'refresher course' even for those who have been licensed a while.

## WHAT IS HF?

HF stands for High Frequency, which is the part of the radio spectrum that concerns us in this book. HF is also known as 'short wave' and, to all intents and purposes, the two terms are interchangeable: high *frequency* equates to short *wavelengths*. Thanks to the reflecting properties of the ionosphere, signals transmitted in this part of the radio spectrum have the remarkable property of travelling great distances, potentially all around the world, in a way that signals transmitted in other parts of the spectrum simply do not. We will discuss this in more detail later in the book.

The simplest form of radio transmission is called a *carrier wave*. By itself a carrier wave contains no information, although it could be pulsed on and off at specific intervals to produce Morse code, for example. Alternatively, modulation (e.g. voice, music or other audio sounds) could be added on to the carrier in different ways to produce an AM or FM transmission – of which more later. But for the moment, let's consider a simple carrier wave containing no other information. A radio carrier wave is in the form of a *sine wave* and is shown in **Fig 1.1**.

If you imagine a wave, it is easy to understand what is meant by wave*length* – see **Fig 1.2**. The wave's *frequency* is no more difficult to grasp: it is simply the number of times the wave makes a complete cycle in one second of time – see **Fig 1.3**. One cycle per second is called one Hertz (Hz), named after the 19th century German physicist Heinrich Hertz. In practice, radio waves make many thousands or millions of cycles in one second, so they are generally measured in thousands of Hertz (kilohertz) or

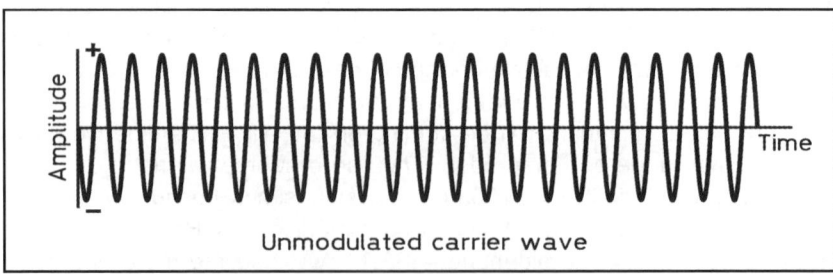

*Fig 1.1: A radio carrier wave, with no modulation.*

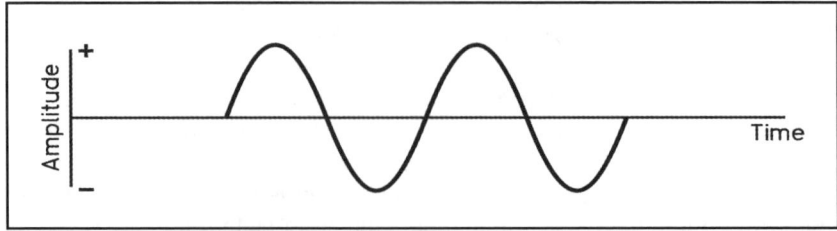

*Fig 1.2: The wavelength is the distance from any one point on the graph to the place where that point is repeated. This could be from peak to the next, or from one trough to the next: the distance will be the same.*

# 1 – WHAT IS HF SSB?

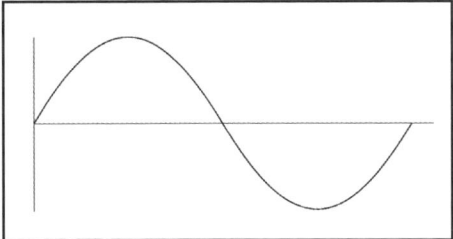

*Fig 1.3: One complete cycle of a radio wave.*

*The German physicist Heinrich Hertz (1857 – 1894), after whom the unit of frequency is named.*

millions of Hertz (Megahertz) and we therefore almost always talk about radio frequencies in kHz or MHz rather than in single Hertz: 1MHz = 1000kHz = 1,000,000Hz.

Note that the abbreviation for kilohertz, kHz, always has a lower case 'k', whereas the abbreviation for Megahertz, MHz, always has an upper case 'M'.

## FREQUENCY & WAVELENGTH

We have already used both the terms 'frequency' and 'wavelength'. The two are related: the higher the frequency, the shorter the wavelength, and vice versa. For example, a radio signal transmitted at a frequency of 5MHz (5000kHz) has a wavelength of 60 metres, while a signal transmitted at a frequency of 20MHz has a wavelength of 15 metres.

There will be almost no mathematics or formulas in this book but, in this case, a very simple formula really *does* help to explain the relationship between frequency and wavelength:

$f = 300 / \lambda$   or   $\lambda = 300 / f$

where $f$ is the frequency in Megahertz (MHz) and $\lambda$ (the Greek letter lambda) is the wavelength in metres.

Where does this 'magic number' of 300 come from? Well, radio waves are just a type of light wave and the velocity of light is approximately 300,000 kilometres per second. We could equally well have written the formula as:

$f = 300,000 / \lambda$   or   $\lambda = 300,000 / f$

where $f$ is still the frequency, but this time in kilohertz (kHz).

Note that wavelengths in metres are often abbreviated with a lower case 'm' (not to be confused with the upper case 'M' in MHz). So 5MHz = 60m, or 20,000kHz = 15m.

**Fig 1.4** shows this relationship between frequency and wavelength in graphical form. The graph can also be used to make approximate calculations from frequency to wavelength and vice versa. If greater accuracy

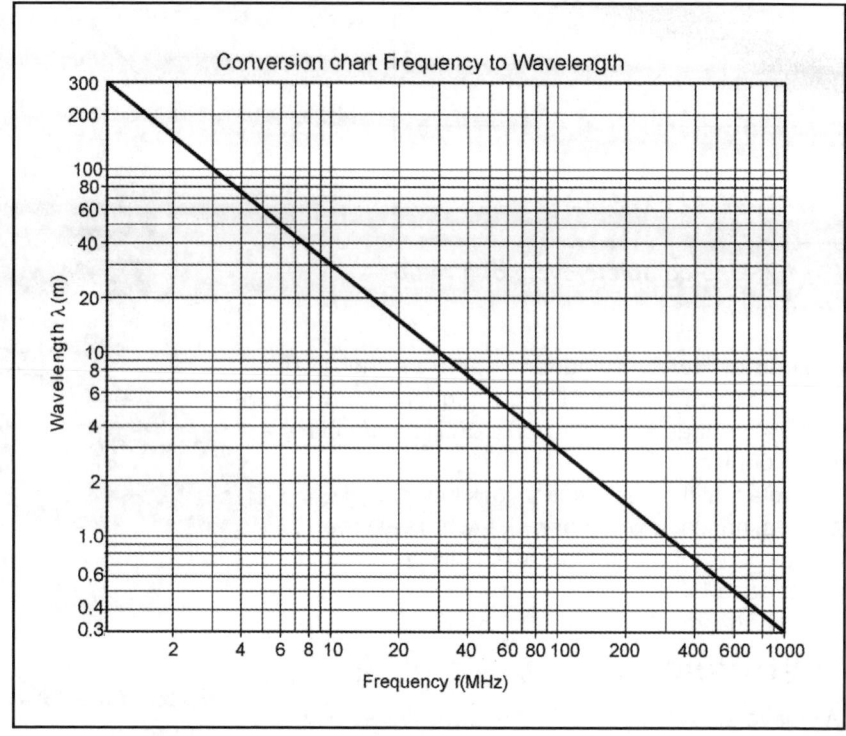

*Fig 1.4: Frequency to wavelength conversion graph.*

is required, even the most basic of pocket calculators will be up to the job: 300,000 divided by 3800 (kHz) = 78.947 metres.

## THE HF SSB AMATEUR BANDS

We started this section with the question "what is HF?" We have explained that the 'H' in HF refers to 'High' Frequencies, and we have explained the relationship between frequency and wavelength. Naturally if there are 'High' Frequencies, there must also be 'Low' Frequencies. Then there are also 'Medium' Frequencies and 'Very High' Frequencies – even 'Ultra High' Frequencies. The latter two of course are better known as 'VHF' and 'UHF', which most people will be familiar with from radio and TV broadcasting. **Table 1.1** shows where HF falls in the radio spectrum.

Below, and especially above (in terms of frequency), the radio spectrum are other types of electromagnetic waves. These include light, X-rays and so on.

There are amateur radio bands both below and above the HF part of the spectrum but this book concentrates on seven of the eight HF amateur radio bands that are available to all three classes of UK licence (SSB is not normally used on the one other band, 10MHz / 30m). **Table 1.2** shows the frequency limits of those bands and also the various ways of referring to each band.

# 1 – WHAT IS HF SSB?

| Spectrum | Abbreviated | Frequency range in kHz | Frequency range in MHz | Wavelengths in metres |
|---|---|---|---|---|
| Very Low Frequency | VLF | 3 – 30kHz | 0.003 – 0.03MHz | 100,000 – 10,000m |
| Low Frequency | LF* | 30 – 300kHz | 0.03 – 0.3MHz | 10,000 – 1000m |
| Medium Frequency | MF** | 300 – 3000kHz | 0.3 – 3MHz | 1000 – 100m |
| High Frequency | HF*** | 3000 – 30,000kHz | 3 – 30MHz | 100 – 10m |
| Very High Frequency | VHF | 30,000 – 300,000kHz | 30 – 300MHz | 10 – 1m |
| Ultra High Frequency | UHF | 300,000 – 3,000,000kHz | 3000 – 3000MHz | 1 – 0.1m |

\* Also known as 'long wave'. ** Also known as 'medium wave'.
\*** Also known as 'short wave'.

Table 1.1: Where HF falls in the radio spectrum.

| Frequency limits (kHz) | Band known as |
|---|---|
| 3500 – 3800 | 3.5MHz or 80m |
| 7000 – 7200 | 7MHz or 40m |
| * 10,100 – 10,150 | 10MHz or 30m |
| 14,000 – 14,350 | 14MHz or 20m |
| 18,068 – 18,168 | 18MHz or 17m |
| 21,000 – 21,450 | 21MHz or 15m |
| 24,890 – 24,990 | 24MHz or 12m |
| 28,000 – 29,700 | 28MHz or 10m |

Table 1.2: The eight HF amateur radio bands available to all three classes of UK amateur licensee. ( * By IARU recommendation, there is no SSB on the 10MHz / 30m band.)

Most commercial amateur radio HF transceivers actually cover more than these eight HF bands. Almost all will also include the 1.8MHz band (1810 – 2000kHz), or 160m. Although most radio amateurs consider this band along with the HF bands, by the definition given above it is not an HF band at all, but rather an MF band. Although it is often used for short-distance contacts, it is the case that longer-distance communications on 160m (DX contacts) do require somewhat different techniques from the HF bands proper, and therefore 160m is somewhat beyond the scope of this book.

There is also another set of frequencies that some UK amateurs can use and which is also in the HF spectrum. Eleven sub-bands around 5MHz (60m) are allocated to amateurs, but only to those with the Full licence: they are not available to Foundation or Intermediate licensees. In the USA, only five 2.8kHz-wide channels are available. In the UK their use was originally intended for research into local and semi-local propagation and for that reason, and because they are not available world-wide or even to all amateurs in the UK, these 5MHz frequencies are not discussed in this book.

Most modern transceivers actually also include the 50MHz (50 – 52MHz) band, or 6 metres, and one or two even include the 70MHz (4m) band. 6m and 4m are of course VHF bands, not HF, and so are also outside the scope of this book, but for those wanting to learn more about those bands the RSGB has published an excellent guide [2].

# HF SSB DX BASICS

When referring to HF frequencies a useful convention, and one used throughout this book, is to use Megahertz when one is referring to either an approximate frequency or a band of frequencies, and to use kilohertz when referring to a specific frequency or a range of frequencies. For example the term '14MHz' is taken to mean the 14MHz band (also known as 20 metres), whereas 14200kHz refers to that specific frequency within the 14MHz band.

## WHAT IS SSB?

SSB stands for 'single side-band'. So, in order to answer the question "what is SSB?" we first really need to know what a 'side-band' (more usually written simply as 'sideband') is, and then we can find out why we might only need a single one of them.

Even the layman, with no knowledge of amateur radio, is familiar with the names of two different types of radio modulation – AM and FM – even if they do not really know what these terms mean. Many people confuse AM and FM with the *frequencies* of operation, a state of affairs not helped by the BBC, which insists on announcing that Radio 2, for example, is broadcast on "88 to 91 FM". In fact, Radio 2 is broadcast from many different transmitter sites around the UK on a number of different frequencies between 88MHz and 91MHz, which are in the very high frequency (VHF) part of the spectrum (see **Table 1.1**). FM stands for 'frequency modulation', which describes the way in which the programme is transmitted or *modulated* on to the carrier wave, but it has nothing to do with the frequency of that carrier signal *per se*.

Radio 4 on 198kHz, a long-wave frequency, on the other hand, is transmitted using a different modulation technique called AM, *amplitude modulation*, as indeed are all the radio stations that broadcast in the long-wave and medium-wave bands.

In fact, AM signals *could* be transmitted on VHF frequencies, and FM signals *could* be transmitted on the medium-wave band: 'AM' and 'FM' refer only to the method (or *mode*) of transmission and not the frequencies on which they are transmitted. FM is capable of providing very high-quality transmissions, but at a price: an FM transmission takes up a lot more of the spectrum (space on the radio dial) than does an AM one, and it is for this reason that, generally, FM transmissions are restricted to very high frequencies, where there is a lot more spectrum available than on the long, medium and short wave bands. Furthermore, VHF signals generally travel shorter distances than medium or short wave signals. This means that Radio 2 on 88 – 91MHz can transmit at exactly the same time and on the same frequencies as, for example, German and Finnish VHF broadcasting stations, without listeners in Germany or Finland receiving interference from Radio 2, or vice versa.

So, FM is an excellent mode of transmission for high-quality broadcasting purposes over the relatively short distances capable on VHF. FM is also used by radio amateurs, again mainly on VHF, e.g. on the 145MHz (2m) band, as well as on UHF, e.g. on the 432MHz (70cm or 70 centimetre) band, for relatively short-distance communications.

## AM TRANSMISSIONS

Let's go back to AM, the mode of transmission used by Radio 4 on long wave, and by BBC Radio 5 Live, as well as all the ILR (commercial) radio stations broadcasting on medium wave. AM is also used by numerous broadcast stations on the short-wave bands, from BBC World Service to the Voice of America and China Radio International.

Take a look at **Fig 1.5**. We have already discussed the carrier in Chapter 2. Its frequency of transmission can be anywhere in the radio spectrum (let's say for example it is on 7100kHz) and it takes up only a tiny amount of that spectrum. As we have discussed, by itself the carrier conveys no information. However, in a process known as *modulating* the carrier, audio information can be added to it. That audio information could be a single-frequency tone at any audio frequency (such as 1000Hz or, for example, a piano's note A above middle C, which is at a frequency of 440Hz),

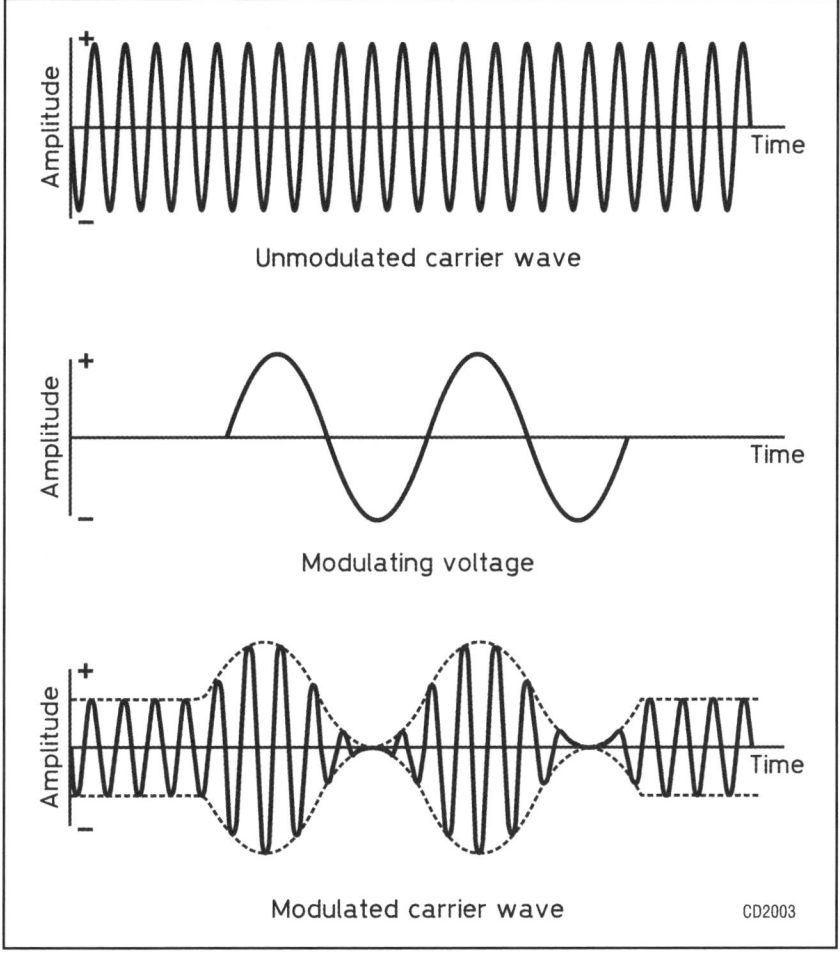

*Fig 1.5: Top, an unmodulated carrier wave. Centre, a single audio tone. Bottom, the carrier wave is being modulated by the audio tone.*

# HF SSB DX BASICS

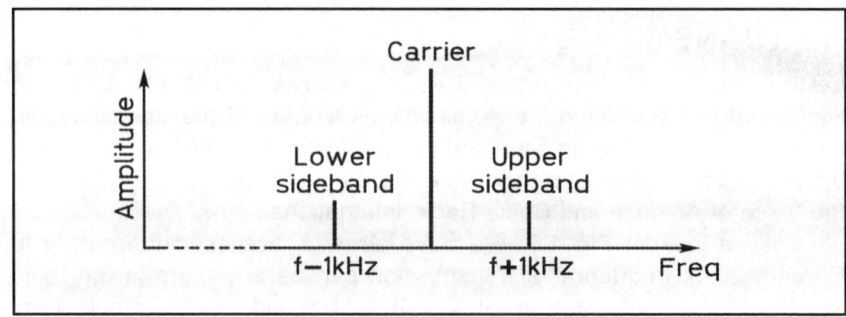

*Fig 1.6: A simple AM transmission: a carrier being modulated by a single tone.*

or it could be speech or in fact any sound at all. **Fig 1.5** shows the carrier being modulated by a single audio tone.

Now look at **Fig 1.6**. This shows exactly the same thing, but this time with respect to the *frequency* of the transmission. In our example the carrier is being transmitted at 7100kHz and in **Fig 1.6** it can be seen that the single tone is at an audio frequency of 1000Hz, or 1kHz. The process of modulating the carrier produces two sidebands, one below and one above the carrier frequency. If the audio tone is 1kHz, the two sidebands will be 1kHz below and 1kHz above the carrier frequency; in the case of our example they are therefore on 7099 and 7101kHz respectively.

Now instead of a single audio tone, consider human speech. The human voice is capable of producing a wide range of audio frequencies (imagine both basso profundo and soprano opera soloists) but the normal everyday speech of men and women falls mainly in the range of 300Hz to 3kHz. Above 3kHz are harmonics and fricatives such as the sibilant 'S' sound, which add 'presence' to the speech, but do not markedly improve its intelligibility. Most AM transmitters are engineered to transmit audio frequencies of up to approximately 3 or 4kHz.

If instead of transmitting a single audio tone, the AM transmitter in the example above were to transmit human speech in the range of 300Hz to 3kHz, the lower sideband would extend from just below the carrier frequency

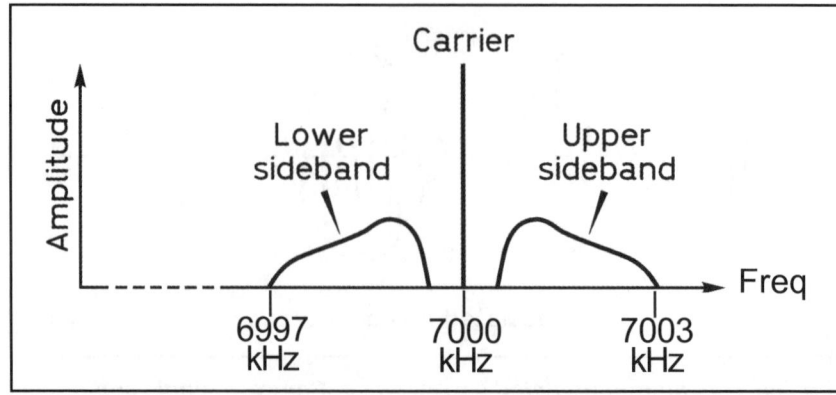

*Fig 1.7: An AM transmission: a carrier being modulated by speech.*

12

# 1 – WHAT IS HF SSB?

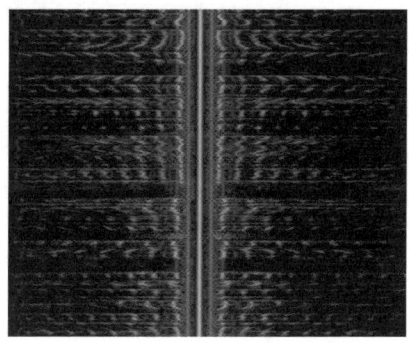

*A spectrogram or 'waterfall display' of an AM transmission. Frequency runs across the picture from left to right, while time runs up and down it. The heavy vertical line in the centre is the carrier and the two sidebands are to the left and right. The three black bands running horizontally across the display, roughly one-quarter, one half and three-quarters of the way down the display, are when the transmission is silent, for example during pauses in speech.*

of 7100kHz down to 7097kHz, while the upper sideband would extend from just above 7100kHz up to 7103kHz, as shown in **Fig 1.7**. The transmission would therefore take up 6kHz of spectrum (7097 to 7103kHz).

Music has a wider audio frequency response than the human voice, from a few Hertz (for the lowest pedal notes on a concert organ) up to around 20kHz, if harmonics are included. An AM transmitter could, theoretically at least, transmit audio frequencies up to 20kHz, but the transmission would then take up 40kHz of spectrum and the limited amount of space available in the long, medium and short wave bands makes this unacceptable: there are simply too many stations trying to fit into the limited amount of spectrum available to make this possible. Being restricted to a bandwidth of 6kHz (audio frequencies of up to 3kHz) means AM transmissions are fine for speech, but sound a little 'muffled' when transmitting music. It is for this reason that music sounds better on an FM transmitter (which is limited to audio frequencies of up to 15kHz).

## FROM AM TO SSB

We have already seen that a single audio tone modulated on to a carrier produces an AM transmission with two sidebands. The tone transmitted by the AM transmitter is the same tone, whether it is on the lower sideband or the upper sideband. Likewise, speech (or music, or any sound) transmitted by an AM transmitter is transmitted both on the lower sideband and, as a 'mirror image', on the upper sideband. This is not only unnecessary, but it is also wasteful: wasteful both of the amount of spectrum required for the transmission, and of the power required to transmit it.

Imagine you have a total of 100 watts of power available for your transmitter. In an AM transmitter some of that power is required for the carrier, some for the lower sideband and some for the upper sideband. But only one of those sidebands is necessary, because identical audio is being transmitted on both of them. If one of the sidebands can be suppressed, more of the available power can be put in to the carrier and the remaining sideband, thus making the signal stronger at the receiver. If the upper sideband is suppressed you are left with a lower sideband or LSB transmission; if the lower sideband is suppressed naturally enough you then have an upper sideband or USB transmission, as shown in **Fig 1.8**. (Those more familiar with computer terminology than that of amateur radio

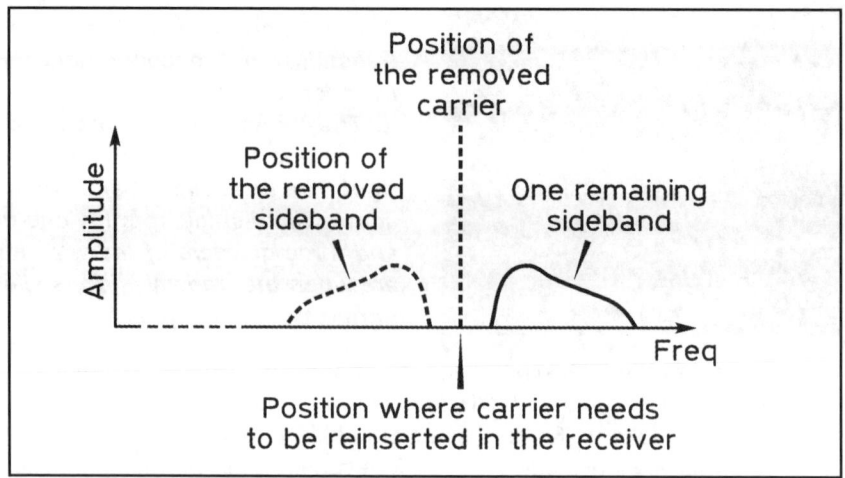

*Fig 1.8: The spectrum of an SSB signal, in this case a USB transmission.*

should get used to the term USB meaning upper sideband. It has been around a lot longer than the Universal Serial Bus, which was only invented in 1996!) As far as the transmission is concerned, it does not matter which sideband is suppressed, as the audio information they contain is identical.

Not only is a significant percentage of power saved by suppressing one of the sidebands, but the *bandwidth* of the signal – the amount of spectrum or space on the radio dial that it takes up – is halved, typically from about 6kHz to about 3kHz. This means that, for a given number of stations operating in a particular band, there is less likely to be interference caused by other stations.

We have said that the carrier, by itself, conveys no information. It does, however, have an important function and that is to provide the receiver with a reference point with which to demodulate the audio signal. In other words, the very existence of the carrier tells the receiver the *precise* frequency of the transmission.

But what if the carrier were also to be suppressed, like the one unnecessary sideband? Two things would happen – and it's good news and bad news. First, the good news: another significant amount of the transmitter's power could be saved or, alternatively, put instead into the one remaining sideband, thus making the signal at the receiver stronger still. Now the bad news. It *does* make the reception of the signal more complex. Just how to tune in an SSB signal on a receiver is covered later.

**Fig 1.8** shows a single sideband transmission with both the lower sideband and the carrier having been suppressed, leaving just the upper sideband remaining.

## HOW SSB IS GENERATED

There are two main ways of generating a single sideband signal; the filter method (shown in **Fig 1.9**) and the phasing method (**Fig 1.10**). Until IF DSP transceivers came along, almost all commercially-made SSB transceivers

# 1 – WHAT IS HF SSB?

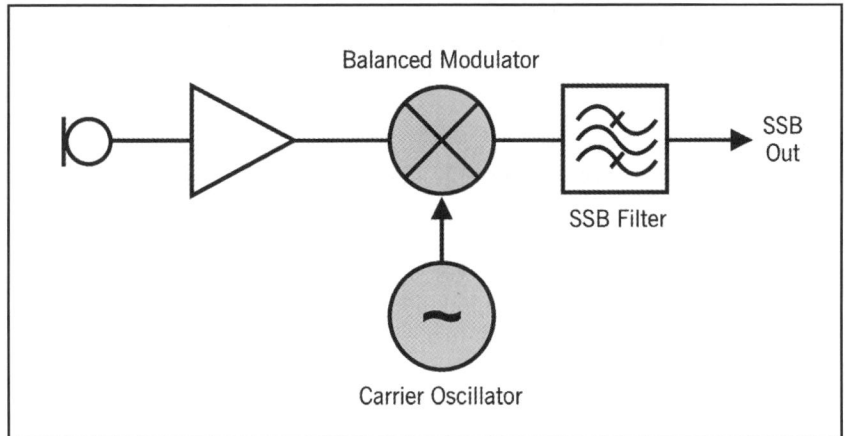

*Fig 1.9: Simplified diagram showing the filter method of generating an SSB signal.*

used the filter method of generating SSB because it was possible to suppress the unwanted sideband to a greater degree than was possible with the phasing method when using analogue techniques. In the filter method, the microphone audio is mixed with the output of the carrier oscillator in a balanced modulator, the output of which is a double sideband signal. This is fed to a crystal or mechanical filter to remove the unwanted sideband as shown in **Fig 1.9**.

The phasing method has tended to be more popular with home-construction enthusiasts because, although it may look more complicated in **Fig 1.10**, it is actually easier to implement than the filter method. Here,

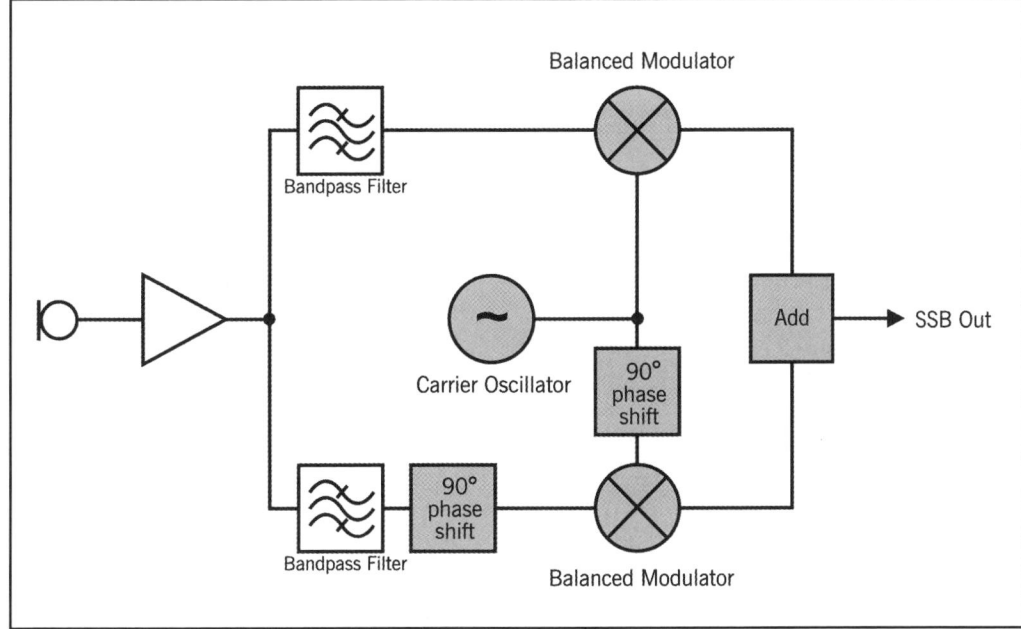

*Fig 1.10: The phasing method of generating an SSB signal.*

the audio from the microphone is first fed to two filters, both of which filter the audio to the final bandwidth that is to be transmitted, such as 300Hz to 2.7kHz. The output of one of the filters is phase shifted by 90° and the outputs of both are passed to two balanced modulators. Both the balanced modulators are also fed with the output of the single carrier oscillator, but the balanced modulator with the phase-shifted audio receives the carrier oscillator's output also phase shifted by 90°. The outputs of the two balanced modulators are then combined: by adding or subtracting the two outputs either an upper sideband or lower sideband signal can be produced. Unfortunately, the level of unwanted sideband suppression can often leave something to be desired in home-made phasing method SSB transmitters.

With the advent of IF DSP transceivers the phasing method has had something of a renaissance, as modern digital techniques make it easier to implement the phasing method in software and allow both the unwanted sideband and the carrier to be suppressed by 60dB or more: at least as good as analogue filtering techniques.

Although the term 'suppressed' is used, it is never possible to suppress the unwanted parts of the signal entirely. A level of suppression of 60dB for both the carrier and the unwanted sideband is considered good. This means, though, that in practice if you are receiving an SSB signal at a signal level of S9+60dB (perfectly possible for a powerful local signal) and the carrier is suppressed by 60dB it will still be S9 at your receiver. It is often possible to hear the carrier of strong SSB signals by tuning slightly lower in frequency for a USB signal, or higher in frequency for an LSB signal. This does not by itself indicate a problem with the transmitter; it is the difference between the peak strength of the wanted sideband and the strength of the carrier that is the issue.

## TUNING IN SSB

Earlier we talked about how an SSB signal is derived from an AM signal and said that the reception of an SSB signal is somewhat more complex than that of an ordinary double sideband (DSB) AM signal. In fact, the receiver circuitry and the actual process of tuning in the signal are both somewhat more complex than that required for AM. We have already seen that an SSB transmitter is also more complex than an AM one.

When tuning-in an ordinary AM signal on a receiver, you do not need to tune the receiver to *exactly* the same frequency as the carrier; you can be up to a couple of kilohertz higher or lower in frequency than the carrier and it does not matter – the signal sounds more or less the same wherever you are listening, provided the signal falls within the passband of the receiver. The same is not true, however, when tuning in a single sideband transmission.

Before the audio can be recovered, the receiver must re-insert the carrier that was suppressed at the transmitter, as was shown in **Fig 1.8**. This is usually done with a circuit called a product detector. (A simpler, though somewhat less effective, way of doing it is to add a beat frequency oscillator – or BFO – circuit to a standard AM receiver.) Not only must the level (the strength) of the re-inserted carrier be correct in order to demodulate the

# 1 – WHAT IS HF SSB?

SSB signal, but also it must be on *exactly* the correct frequency.

Fortunately for us amateurs the tolerance required when tuning in speech, especially speech of 'communications quality', is much less than that required for music. For ordinary speech to sound 'natural' it needs to be tuned in to within about 50Hz (0.05kHz) of the suppressed carrier frequency and, with today's modern and stable receivers and transceivers, that is not too difficult to achieve (although it may take a little practice).

Beginners to amateur radio, who may well be used to tuning in AM and FM radio broadcasts, often find it difficult to tune in an SSB transmission at first. But the knack soon comes. How is it done?

All transceivers, and almost all receivers – at least 'communications receivers', those intended for reception of transmissions other than standard AM or FM broadcasts – have a *mode* switch (or perhaps a series of push buttons) on the front panel to select the appropriate mode. Among the modes available (including CW, AM, probably FM and perhaps RTTY and / or Packet) there will be switch positions or buttons marked 'LSB' and 'USB'.

Virtually all properly-licensed *non*-amateur SSB transmissions are on USB (no doubt there are a few exceptions, but I am not aware of any). In the case of *amateur* SSB transmissions, though, there is a convention that dates back to the early days of SSB communications that has LSB being used on 40 and 80m (and 160m), and USB on the other bands. There is nothing sacrosanct about this: amateurs may, if they wish, transmit USB on 40 and 80m and LSB on the higher-frequency bands (although they might not make many contacts if they did!)

It is impossible to receive a USB transmission if the receiver is switched to LSB and vice versa or, to be pedantic, it is possible to *receive* the signal but it will be impossible to understand it, because it will sound completely 'scrambled'. So, before tuning in the SSB signal it is important to know whether the transmission is on lower sideband or upper sideband.

You can assume that nearly 100% of radio amateurs will stick to the convention of operating LSB on 40 and 80 metres (and also 160 metres), and USB elsewhere. Select the appropriate sideband and tune across the band. You will note that as you tune across an SSB signal, the voice will sound distorted and either unnaturally high-pitched or unnaturally low-pitched. Tune across the signal slowly – *very* slowly if necessary – and as you do so you will notice that the pitch of the voice changes. At first, it is easy to 'overshoot' and tune beyond the correct point, in which case obviously you must make an adjustment by turning the tuning knob in the opposite direction by a small amount. At one point between too high-pitched and too low-

*The mode buttons on a modern transceiver. Be sure to select the correct sideband before attempting to tune in an SSB signal.*

pitched it will be 'spot on'.

At first the process of tuning in an SSB signal may seem quite random, but it will soon be realised that all the signals on the band tune 'the same way'. *Which* way is dependent on which sideband is being transmitted. On USB if you start tuning from the bottom of the band towards the top end, i.e. if you are tuning higher in frequency, as a signal comes into the passband of the receiver it will first sound too high-pitched and, if you have gone 'too far', it will sound too low-pitched. On LSB it is the other way round.

Although tuning-in SSB signals very soon becomes second nature, when starting out many people find it easiest always to tune across a band in the same direction. In that way when a signal is not quite correctly tuned-in, the operator instinctively knows which way to turn the tuning knob to make the correct adjustment. This is not a bad habit to get into and even now, some 45 years after listening to my first SSB signals, I usually find myself tuning 'from high to low' on 80 and 40 metres, and 'from low to high' on the higher-frequency bands. For example on 80 metres I start at 3800kHz and tune lower in frequency, while on 20 metres I start at about 14110kHz and tune higher in frequency. In both cases SSB signals coming into the receiver's passband start by sounding too high-pitched and go lower in pitch as the signal is tuned in.

This may sound complex but it is one of those things that is much more difficult and long-winded to describe in words than actually to do in practice. After a bit of practice using a particular receiver or transceiver you will find that not only do you know instinctively which way to turn the tuning knob to tune in a signal correctly, but also you will know how *much* to turn it – even to the extent of being able to judge how far to turn the tuning knob between short transmissions, when the signal is not actually on the air, in order for the signal to sound natural when the transmission recommences.

## TUNING RATE

One final tip for tuning-in SSB signals. With many transceivers and receivers it is possible to adjust the tuning *rate* – the amount the equipment changes in frequency as the tuning knob is turned. It is important to ensure that the best rate is selected before attempting to tune in any signals. Too fast and it is all too easy to 'overshoot' and so it becomes very difficult to tune in the signal to exactly the correct spot. Too slow and tuning in the signal becomes laborious and, in 'quick-fire' operating conditions such as during a DXpedition or a contest, the wanted station is likely to have ended its transmission before you have had a chance to tune it in correctly, no matter how fast you try to turn the tuning knob.

So what is the 'best' tuning rate? To some extent this is a matter of personal choice. For example, I like a tuning rate of about 10kHz per 360° rotation of the tuning knob, while some operators may prefer as little as 1kHz per full rotation. I have noticed that those who are primarily CW (Morse code) operators, and only venture on to SSB occasionally, generally prefer a much slower tuning rate than those of us who are primarily or exclusively SSB operators. On the other hand, 100kHz per revolution is *way* too fast for

# 1 – WHAT IS HF SSB?

*The AOR ARD9800 'Fast Radio Modem' for DV SSB.*

SSB (though it might be suitable for AM or for rapidly moving from one end of the band to the other). Anywhere between about 5 and 15kHz per 360° rotation is probably about right for SSB reception.

## SSB DIGITAL VOICE

A quick word about Digital Voice on SSB. Here we do not mean conventional SSB generated by digital techniques, such as those described earlier in the chapter, but rather a mode in which the voice is first digitised and then transmitted over an SSB signal and bandwidth. It is therefore better described as Digital Voice, or 'DV', than as digital SSB. DV has been around in one form or another since 1999 but has yet to become a 'mainstream' or popular mode (in contrast to D-Star and other forms of digital voice used on VHF).

The Japanese company AOR released the ARD9800 'Fast Radio Modem', the first commercially-available SSB digital voice equipment, in 2004. It is based on earlier pioneering work carried out by Charles Brain, G4GUO, and Andy Talbot, G4JNT, and uses an open protocol developed by G4GUO. The ARD9800 is a stand-alone unit, which simply connects to the microphone input and speaker output connections of an SSB transceiver and a suitable 12V DC supply. With an optional memory module, it can also send photos or still video images rather like SSTV, and data at a speed of 3600bps.

The IARU Region 1 HF band plans list Digital Voice centres of activity at 3630, 7070, 14130, 18150, 21180, 24960 and 28330kHz.

DV provides improved speech quality compared with conventional analogue SSB, equivalent to VHF narrow-band FM quality, but at the price of requiring a signal-to-noise ratio of about 25dB in order to work effectively. It is therefore *not* a weak-signal mode and so, unlike conventional SSB, DV is quite unsuitable for DX working.

## REFERENCES

[1] *The Complete DX'er*, Bob Locher, W9KNI, 3rd Edition, Idiom Press, 2003.
[2] *Six and Four* ('The Complete Guide to 50 & 70MHz Amateur Radio'), Don Field, G3XTT, RSGB 2013, available from www.rsgbshop.org

# 2   What is DX?

THIS BOOK IS CALLED *HF SSB DX Basics* but although we have defined HF and SSB we have not yet defined precisely what is meant by the term 'DX'. That is not as easy as it may sound, as we shall see. Perhaps the best definition was coined by the late Californian amateur, writer and humorist Hugh Cassidy, WA6AUD, who simply stated "DX IS". In other words, and to paraphrase Lewis Carroll, DX can mean whatever you want it to mean. Some sort of definition would be helpful, though, so here goes . . .

On the VHF / UHF bands 'DX' equates to distance. The further the contact, the greater the DX. To some extent the same is true on 160m, where it can be difficult to make any QSOs outside your own continent, but that's not really the case on the HF bands from 10 to 80m.

When a station puts out a "CQ DX" call on any of the HF bands it is generally understood to mean that the station is looking for contacts outside their own continent. It would therefore be considered legitimate for a US station to reply to a "CQ DX" call from someone in the UK, for example. Nevertheless, this definition does not really work either: most experienced amateurs would not really consider a contact with the Canary Islands (considered to be in Africa) or even New York to be 'real DX'.

Distance isn't the issue, it's more to do with the 'rarity' of the station. From the UK, although any contact with North America is outside one's own continent, there are so many amateurs in the US, many of whom have large stations and antenna systems, that the east coast of the US is not considered to be 'DX' by most HF operators. So a contact with Florida wouldn't be DX, yet contacts with the Bahamas (C6A), the Turks and Caicos Islands (VP5) or Cuba (CO) – none far from the coast of Florida – *would* be considered DX, even though the distances involved are similar.

## WHAT CONSTITUTES 'RARE' DX?

So now we have defined 'DX' – sort of. It is now clear that there is what might be called 'ordinary DX' and 'rare DX'. For many years the American publication *The DX Magazine* [1] has run a survey in September – October each year to determine which are the 'Most Wanted' (and therefore the 'rarest') amateur radio 'countries'. Until recently *The DX Magazine's* survey provided the only such available data, but the massive database of *Club*

*Log* [2] allows anyone to generate a 'Most Wanted' list from the 315 *million* log entries (as of November 2015) that have been uploaded to it. (More about *Club Log* later in this chapter.)

**Table 2.1** shows the result of an interrogation of the *Club Log* database in early November 2015. The data shown in the table are from all (world-wide) logs, using all bands, but on SSB only. It is also possible to determine which are the 'Most Wanted' entities in any particular continent or part of a continent, on any particular mode and on any particular band.

Although the *Club Log* Most Wanted list will change from time to time (for example as a result of an activation by a major DXpedition) some factors remain constant. DXCC entities are rare for one of two reasons. The South Sandwich Islands and Heard Island (ranked 2nd and 3rd respectively) are rare because they lie deep in the Southern Ocean, and any expedition is hugely expensive to mount. North Korea (ranked 1st) is relatively easy to reach, has an international airport and modern hotels, but licensing is very difficult for political reasons.

| Rank | Prefix | Entity Name |
|---|---|---|
| 1 | P5 | DPRK (North Korea) |
| 2 | VP8S | South Sandwich Is |
| 3 | VK0H | Heard Island |
| 4 | FT5J | Juan de Nova, Europa |
| 5 | 3Y/B | Bouvet Island |
| 6 | FT5W | Crozet Island |
| 7 | KH5 | Palmyra & Jarvis Is |
| 8 | BS7H | Scarborough Reef |
| 9 | KH5K | Kingman Reef |
| 10 | VP8G | South Georgia Island |
| 11 | KH1 | Baker & Howland Is |
| 12 | BV9P | Pratas Island |
| 13 | VK0M | Macquarie Island |
| 14 | CE0X | San Felix Is |
| 15 | FT5X | Kerguelen Island |
| 16 | FK/C | Chesterfield Is |
| 17 | KH7K | Kure Island |
| 18 | KH3 | Johnston Island |
| 19 | SV/A | Mount Athos |
| 20 | JD1 | Minami Torishima |

*Table 2.1: Club Log 'Most Wanted' list (November 2015), see text.*

## A BIT OF HISTORY

Radio amateurs have always been DXers. Indeed, a case can be made that Guglielmo Marconi was the world's first DXer, as he was always striving to make transmissions over greater and greater distances. Two-way communications across the Atlantic were achieved by radio amateurs in a series of tests in 1923. Then, on 18 October 1924, a two-way contact was made by Cecil Goyder, 2SZ, a pupil at Mill Hill School in North London, and Frank Bell, 4AA, a sheep farmer on the South Island of New Zealand. World-wide DX had arrived and it was not long before radio amateurs were comparing not only how *far* their signals had travelled, but also the number of countries they had contacted.

But what constituted a 'country'? This question was tackled by Clinton B DeSoto, W1CBD, in an article entitled 'How to Count Countries Worked, A New DX Scoring System' that was published in the ARRL members' journal *QST* in October 1935 [3]. DeSoto suggested that

*Clinton B DeSoto, W1CBD, author of the 1935 article 'How to Count Countries Worked, A New DX Scoring System'.*

# HF SSB DX BASICS

> ## THE 'DELETED LIST'
>
> Once an entity has been added to the DXCC List, the List remains unchanged until that entity no longer satisfies the criteria *under which it was originally added*, at which time it is moved to the 'Deleted List'.
>
> Many entities have come and gone over the years. For example, the former East Germany (Y2) ceased to be a separate entity upon the reunification of Germany and so was moved to the Deleted List. Czechoslovakia, on the other hand, is also now on the Deleted List but was replaced by two new entities, the Czech Republic (OK / OL), and Slovakia (the Slovak Republic, OM).

"The basic rule is simple and direct: Each discrete geographical or political entity is considered to be a country." Two years later the ARRL published the first 'DXCC List' – the definitive list of amateur radio 'countries'. It was therefore never the case that all DXCC 'countries' would be countries in the traditional sense of the word. The criteria for inclusion on the DXCC List have been changed a number of times over the years but DeSoto's basic rule still exists.

Originally, amateurs were happy to work each country just once, regardless of the band or mode. However, as amateur radio activity increased after WWII with many more countries becoming active on the air higher country totals became easier to achieve and amateurs started to look for new challenges. In 1969 the DXCC programme was extended with the introduction of the 5-band DXCC award (in those days there were no 30, 17 or 12m bands, and 160m was excluded on the grounds that many countries still had no 160m allocation). As a result, amateurs who had high country totals on, say, 20m or 15m started to chase countries on 80 and 40m.

Awards programmes, based on the numbers of entities or islands contacted, have become a way of measuring one's DXing success. DXCC is the most popular, followed by the RSGB's Islands On The Air programme, so let's take a look at both in more detail.

## DXCC TODAY

Today's DXCC List **[4]** is comprised of 340 entities, the most recent addition being that of South Sudan (originally ST0, now Z8) after it gained its independence in 2011.

These days, in addition to the basic 'all-band' DXCC award, the ARRL offers separate DXCC awards for 11 bands, from 160 to 2m. There is also a 'DX Challenge' programme which includes the 10 bands from 160m to 6m. Many DXers now look for contacts with any given DXCC entity on all 10 bands, and some also on SSB, CW and data modes! The thought of trying to contact each of the 340 DXCC entities on 10 bands *and* three modes (which would require over 10,000 contacts) might be enough to put anyone off before they even start but fortunately you don't have to start that way. Set yourself a reasonable goal, such as working DXCC entities on your favourite mode such as SSB and on any band or combination of bands.

*QSL from the first operation from South Sudan following its independence in 2011.*

# 2 — WHAT IS DX?

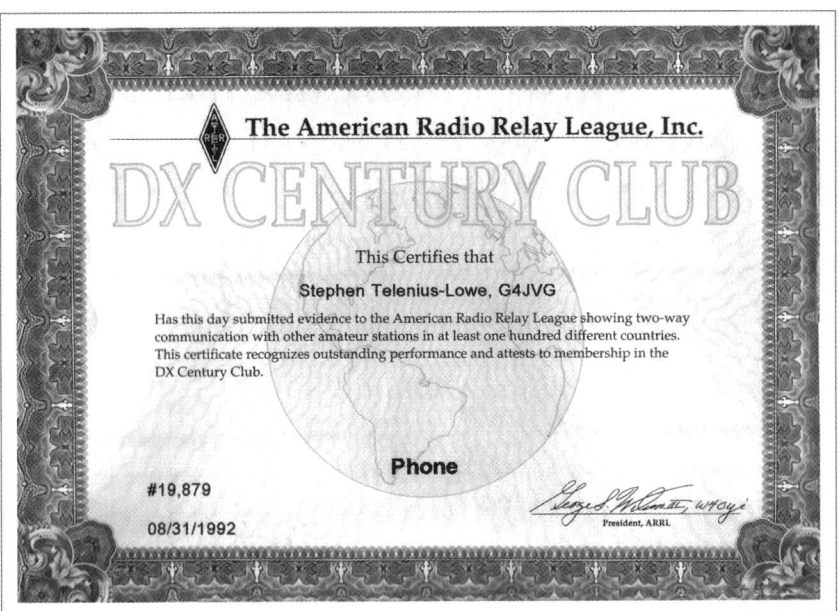

*A basic (all-band) DXCC certificate, endorsed for Phone (SSB) only contacts, awarded to the author back in 1992.*

The basic DXCC certificate is awarded to those who can show confirmations – either QSL cards or matches on the ARRL's Logbook of The World (LoTW) system (of which more later) – from a minimum of 100 entities on the current DXCC List.

All contacts must be made using callsigns issued to the same licensee. Therefore you can 'feed' a single DXCC award from several different callsigns. For example if you start DXCC as a Foundation licensee, progress through the Intermediate licence and eventually upgrade to a Full licence, all the contacts made with the M6 and 2E0 callsigns can be counted towards your M0 DXCC award. Likewise in the USA if you change from the issued 2 x 3 callsign to a shorter 'vanity' callsign contacts made with the original callsign still count. However, you must make all the contacts *from within the same DXCC entity* so, if you start your DXCC in England but then move to Scotland, you must start DXCC all over again.

## LOGBOOK OF THE WORLD (LOTW)

Logbook of The World (LoTW) is an online service provided by the ARRL that allows all amateurs to upload electronic logs to a massive central database. Having done so, DXers can then view their submitted QSOs and check which of their contacts 'match' those of the millions of other QSOs uploaded to LoTW. If the information in a submitted QSO matches the information submitted to LoTW by your QSO partner, the LoTW accounts of both you and your QSO partner will show the submitted QSO as 'confirmed'. With your LoTW account, you can then

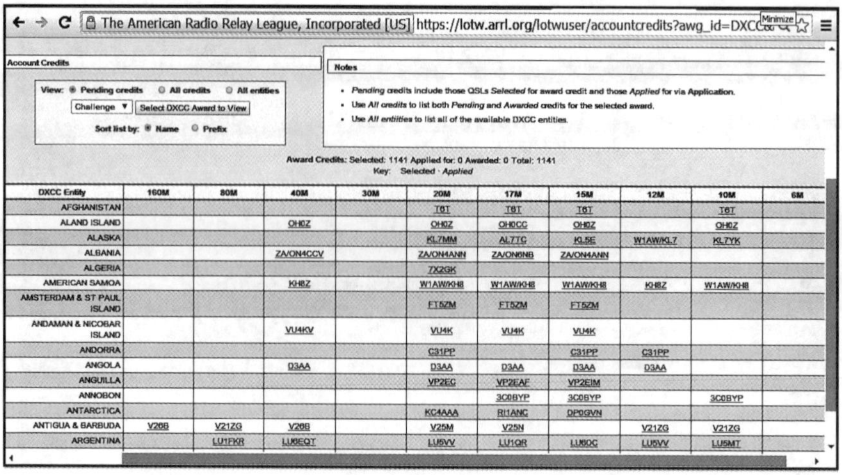


| DXCC Entity | 160M | 80M | 40M | 30M | 20M | 17M | 15M | 12M | 10M | 6M |
|---|---|---|---|---|---|---|---|---|---|---|
| AFGHANISTAN | | | | | T6T | T6T | T6T | | T6T | |
| ALAND ISLAND | | | OH0Z | | OH0Z | OH0CC | OH0Z | | OH0Z | |
| ALASKA | | | | | KL7MM | AL7TC | KL5E | W1AW/KL7 | KL7YK | |
| ALBANIA | | | ZA/ON4CCV | | ZA/ON4ANN | ZA/ON6NB | ZA/ON4ANN | | | |
| ALGERIA | | | | | 7X2GK | | | | | |
| AMERICAN SAMOA | | | KH8Z | | W1AW/KH8 | W1AW/KH8 | W1AW/KH8 | KH8Z | W1AW/KH8 | |
| AMSTERDAM & ST PAUL ISLAND | | | | | FT5ZM | FT5ZM | FT5ZM | | | |
| ANDAMAN & NICOBAR ISLAND | | | VU4KV | | VU4K | VU4K | VU4K | | | |
| ANDORRA | | | | | C31PP | | C31PP | C31PP | | |
| ANGOLA | | | D3AA | | D3AA | D3AA | D3AA | D3AA | | |
| ANGUILLA | | | | | VP2EC | VP2EAF | VP2EIM | | | |
| ANNOBON | | | | | | 3C0BYP | 3C0BYP | | 3C0BYP | |
| ANTARCTICA | | | | | KC4AAA | RI1ANC | DP0GVN | | | |
| ANTIGUA & BARBUDA | V26B | V21ZG | V26B | | V25M | V25N | | V21ZG | V21ZG | |
| ARGENTINA | | LU1FKR | LU6EQT | | LU5VV | LU1QR | LU6OC | LU5VV | LU5MT | |

***Fig 2.1: Extract from PJ4DX's LoTW account.***

submit a confirmed QSO for credit to your DXCC award (as well as Worked All States and the WPX – Worked All Prefixes – award).

Full information about LoTW and how to register with it can be found of the ARRL website **[5]**. Membership of ARRL is not a requirement and there is no fee for using LoTW although there is a charge made if and when you submit confirmed QSOs to credit an award application.

Having achieved DXCC certificates in five different countries where I have been resident, by using only traditional paper QSL cards, when I moved to Bonaire in 2013 and started DXCC all over again – for the sixth time – I took the decision *not* to request any further QSLs (though my QSL manager will send my card to those who request one from me). Instead I am working towards DXCC entirely through the LoTW programme. To give an indication of how popular LoTW has become in recent years, in the two years since November 2013 I have worked 280 DXCC entities and 239 of them have been confirmed on LoTW (see **Fig 2.1**).

## CLUB LOG

*Club Log* **[2]** is a great and fairly new tool for all amateurs that is of special interest to DXers. Established by Michael Wells, G7VJR, and maintained by him and a small team of volunteer helpers, *Club Log* is a web-based application that analyses logs submitted by amateurs all over the world.

Without actually using *Club Log* it is almost impossible to envisage just how much useful data it can provide for you. Among the numerous features are personal DXCC reports and analysis of your log, a timeline of your activity with DXCC entities worked each year and band and mode information, access to propagation predictions using everyone's logs, and OQRS (Online QSL Request Service) to make direct and bureau QSLing faster, easier and cheaper – and much more besides.

At the time of writing (early November 2015) over 315,000,000 QSOs had been uploaded to *Club Log*, with an average of around 1100 new logs

# 2 — WHAT IS DX?

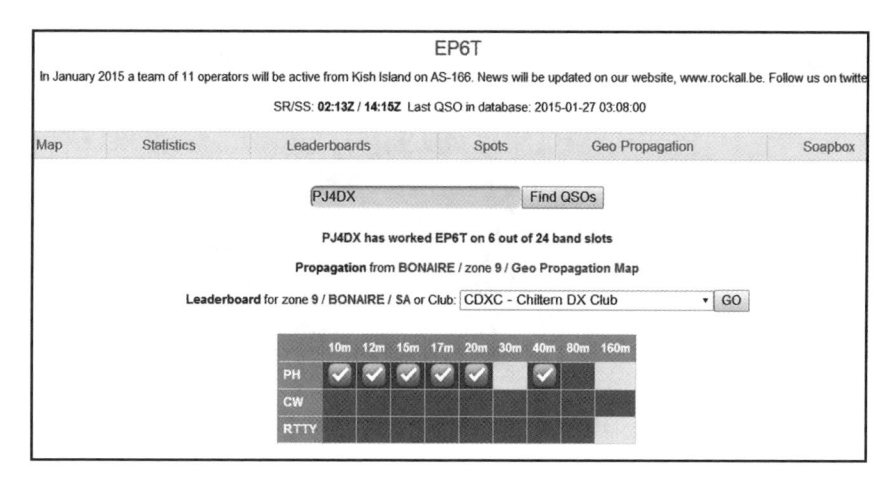

*Fig 2.2: Club Log DXpedition log check.*

being uploaded each day. Everyone is requested to upload their log, no matter how big or small.

Many DXpeditions now upload their logs to *Club Log* during the DXpedition itself (assuming they are in a location that has Internet access), and this allows DXers to check that they are in the DXpedition log, and on which bands and modes (see **Fig 2.2**). In the example shown above, the screen shows that PJ4DX worked the EP6T (Iran, 2015) DXpedition on SSB on six bands, 10 – 40m. It also shows that EP6T was also active on 80m SSB, all bands 10 – 160m on CW, and all bands 10 – 80m on RTTY, but I made no contacts on any of those band / mode combinations. Much more analysis can be made from this page alone by clicking on the 'Propagation' and 'Leaderboard' buttons on the screen.

*Club Log* is emphatically not a rival to the ARRL's LoTW as it does not support applications for DXCC or other awards. Rather it is a means of providing and analysing huge amounts of data. The 'Most Wanted' list shown in **Table 2.1** was derived by interrogating the *Club Log* database.

## ISLANDS ON THE AIR (IOTA)

So far we have only discussed the ARRL DXCC award. While that is certainly the most popular DX-orientated on-air activity programme, it is by no means the only one. The other major player on the world DX stage is the RSGB's Islands On The Air (IOTA) programme. IOTA is now second only to DXCC in terms of on-air activity associated with the programme.

IOTA celebrated its 50th anniversary in July 2014 and in that time the programme has grown to 2500 active 'island chasers' and approximately 15,000 more casual participants.

The basic building block for IOTA is the IOTA Group, of which there are about 1200, with varying numbers of qualifying islands in each. Each group activated is issued with an IOTA reference number, for example EU-005 for Great Britain. The *IOTA Directory* **[6]** provides a full listing of all the IOTA groups, together with the names of some 15,000 qualifying islands.

*The basic IOTA 100 'Islands of the World' certificate, awarded to the author for contacts made from East Malaysia (9M6).*

There is a wide range of separate certificates and awards available for island chasers. IOTA also has an Annual Listing and an Honour Roll. The Annual Listing is a list of the callsigns of stations with a checked score of 100 or more IOTA groups but less than the qualifying threshold for entry into the Honour Roll. The Honour Roll is a list of the callsigns of stations with a checked score equalling or exceeding 50% of the total of numbered IOTA groups, excluding those with provisional numbers, at the time of preparation. The Annual Listing and Honour Roll are published on the RSGB IOTA website **[7]** and in the *RSGB Yearbook* **[8]**. All you need to enable you to participate in the Annual Listing or Honour Roll is the basic IOTA 100 certificate.

In October 2015, the RSGB announced that a new organisation called the IOTA Foundation (IOTAF) will manage the IOTA programme in partnership with the RSGB. One major task for the new organisation will be to develop a new online island credit submission system akin to the ARRL's Logbook of the World. This is due to be completed in 2017.

## REFERENCES

[1] *The DX Magazine*: www.dxpub.com
[2] *Club Log*: www.clublog.org
[3] 'How to Count Countries Worked: A New DX Scoring System', Clinton B DeSoto, W1CBD: www.arrl.org/desoto
[4] The DXCC List: www.arrl.org/country-lists-prefixes (click on 'Go Now')
[5] LoTW on the ARRL website: www.arrl.org/logbook-of-the-world
[6] *IOTA Directory*, Roger Balister, G3KMA, 50th anniversary edition (2014), available from www.rsgbshop.org
[7] RSGB IOTA website: www.rsgbiota.org
[8] *RSGB Yearbook*, available from www.rsgbshop.org

# 3 Planning Your Antenna

THERE IS A TENDENCY for newcomers to buy a commercially-made HF antenna at the same time as they buy their HF transceiver. After all, you need both a transceiver and an antenna in order to start making some contacts. However, I would urge you not to do this. Even those of us who might best be described as 'non-technical' hams, and for whom building an HF SSB transceiver would be about as likely as building our own jet aircraft in our garage, can nevertheless still easily build very effective HF antennas.

Those who *do* buy a commercial HF antenna invariably buy something that they think they can 'get away with' in their garden, in other words something low profile such as a small vertical or the ubiquitous G5RV. But these antennas are not necessarily the most suitable for many people's circumstances. Sometimes something that can easily be made at home will not only cost a fraction of the amount of a commercially-made antenna but also actually be far more effective.

We HF operators have a distinct advantage over VHF / UHF enthusiasts when it comes to building antennas. Because the wavelengths we are dealing with are much longer, the permissible tolerances when making antennas are much greater. When building an antenna for 80m, it is quite normal to have to trim it by a foot or two (30 – 60cm). Someone building a 70cm beam has to be accurate to within a few millimetres or a tiny fraction of an inch.

This chapter is entitled *planning* your antenna, though, and not 'building your antenna', because its main intention is to give you enough information to enable you to make a decision about what *sort* of antenna would be suitable for your particular set of circumstances. There are many excellent antenna books available, some of which are listed as references at the end of this chapter (e.g. **[1]**, **[2]**, **[3]**) which will provide you with far more information about the theory and practice of HF antennas. Nevertheless, a few designs of HF antennas are given, if only to illustrate how simple many of these are to make.

## HORIZONTAL OR VERTICAL?

First, though, a few words about the differences between horizontal and vertical antennas. Generally, an antenna that is erected in the horizontal plane will have horizontal polarisation and one in the vertical plane vertical polarisation. Actually, whether it is horizontally or vertically polarised is not important as the polarisation will be mixed up in the ionosphere and the signal returning to earth is likely to have components of both. Either horizontal or vertical antennas can be used, but each have their own distinct properties.

A vertically-polarised antenna will tend to have a low vertical angle of radiation (this is the angle between the major lobe of the antenna, where the maximum radiation occurs, and the horizon) and, for the long-distance working that is of most interest to DXers, the lower the angle of radiation the better. There are exceptions: a three-quarter wavelength long vertical antenna has a high angle of radiation, for example. However, almost all practical vertical antennas are less than 5/8-wave long and have a low angle of radiation.

A horizontally-polarised antenna's angle of radiation is generally determined by its height above ground. The higher the antenna, the lower the angle of radiation. And herein lies a problem for horizontally-polarised antennas, particularly on the lower-frequency bands. In order for a horizontal dipole (for example) to have a good low angle of radiation, it must be mounted *at least* a half-wavelength above ground. In the case of 7MHz this means around 20 metres (65ft) high, or in the case of 3.8MHz it means 39 metres or 128ft high. Clearly these are heights that are not realisable unless you have access to some big towers or perhaps can put up an antenna between two high-rise apartment blocks!

This does not mean that an 80m dipole less than 39m high won't work at all, though. On the contrary, it will actually work very well for some purposes at more or less any height. What is does mean, though, is that the main lobe of the dipole's radiation will be at a relatively high angle. This is actually *good* for short and medium-distance communications, e.g. for working around the UK and most of Europe, but it will be lacking for long-distance communications, such as into South America, Southern Africa, East and South-East Asia and particularly the Pacific.

To work into these parts of the world a lower angle of radiation is necessary, and on the lower-frequency bands this means either a very high horizontal antenna or a vertically-polarised one (even if the latter is ground mounted). On 14MHz and above there is less of a problem: a half-wave on 20m is only 10m (33ft), so it is not out of the question to raise a simple horizontal antenna such as a dipole to this height or greater.

Many DXers, whose main interest is in working as many countries as possible around the world, tend to use horizontally-polarised antennas (such as a Yagi beam) around 40ft or more high for the bands between 14 and 28MHz, and vertically-polarised antennas (such as quarter-wave verticals) on 3.8 and 7MHz.

## WHAT ANTENNA?

The whole point of this chapter is to give some pointers as to which antenna or antennas might be best for DXing on the bands between 80m and 10m. First, though, we need to make some assumptions. It is assumed that most people will not start out with monoband Yagi beams on each of the higher-frequency bands, with perhaps a 4-Square array of four quarter-wave

## 3 — PLANNING YOUR ANTENNA

verticals on 80m. If you have that sort of wherewithal, you need read no further in this chapter! But for the average person, how to start – bearing in mind that the intention is to work DX, and on SSB too?

Operators using CW and especially digital modes such as PSK31 can often get away with using an inferior antenna, as signals can still be read at lower levels and in the presence of noise to a much greater degree than they can with SSB. Furthermore, Foundation or Intermediate licensees are restricted to using 10W or 50W power output respectively, and so their signal will already be several decibels down on someone using 100W or more. The important thing, therefore, is to use the most *efficient* antenna possible, given your own particular circumstances. If you live on a farm out in the country you will have few restrictions in this respect. For the average amateur living in suburbia it is a matter of using compromise antennas that the neighbours will allow them to 'get away with' – that phrase again. What you should *not* do, though, is put up the smallest or least conspicuous antenna you can, as it probably would not be very effective, especially for DX working. I hope this chapter will provide a few ideas.

## DIPOLES

The humble dipole should not be ignored as a practical DX antenna on the HF bands. It is a half-wavelength long and consists of two wires of identical length, 'fed' in the centre, usually with coaxial cable (see **Fig 3.1**). A dipole is extremely simple to make, it is very light in weight, by using thin wire it can be made practically invisible (so is suitable for those who may have intolerant neighbours), by using the house and a suitable tree as supports no mast is required and therefore there is unlikely to be a requirement for planning permission and, last but certainly not least, the dipole is actually a very efficient antenna. The only two downsides are that (with one or two exceptions) it is a single-band antenna (so if you wanted to operate on, say, seven HF bands you would ideally need seven separate dipoles), and that it needs to be mounted up high if it is to be really effective as a DX antenna.

The dipole theoretically has a so-called 'figure-of-eight' horizontal radiation pattern (see **Fig 3.2**), with maximum radiation broadside to the direction of the wires and little or no radiation off the ends of the wires. Most antenna books will tell you to ensure

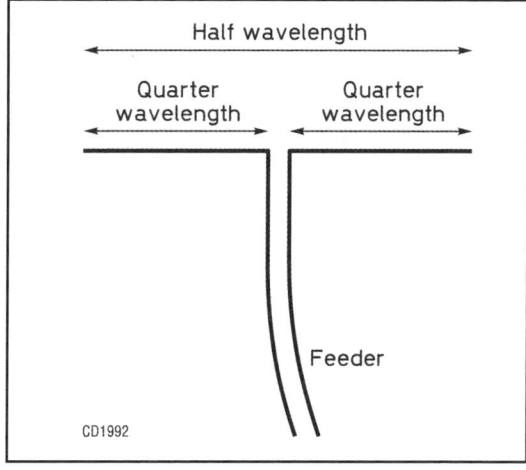

*Fig 3.1: The basic half-wave dipole. The feeder would normally be coaxial cable.*

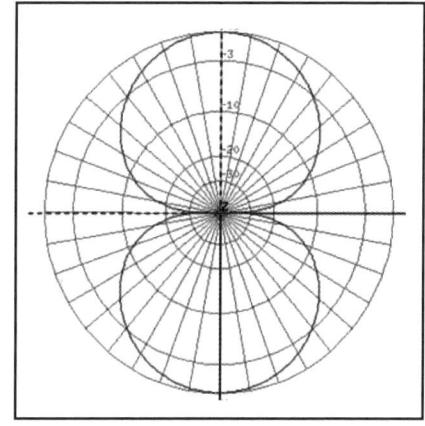

*Fig 3.2: The dipole's theoretical radiation pattern, a so-called 'figure-of-eight' pattern.*

29

# HF SSB DX BASICS

| Band | Frequency | Length | |
|------|-----------|--------|---|
| 80m | 3785kHz | 37.64m | 123ft 6in |
| 40m | 7150kHz | 19.89m | 65ft 3in |
| 20m | 14200kHz | 10.00m | 32ft 9in |
| 17m | 18150kHz | 7.82m | 25ft 8in |
| 15m | 21250kHz | 6.70m | 22ft 0in |
| 12m | 24950kHz | 5.69m | 18ft 8in |
| 10m | 28500kHz | 5.00m | 16ft 4in |

*Table 3.1: Approximate 'real life' lengths of half-wave dipoles for the SSB DX band segments.*

that you orientate it so that the maximum radiation is in the directions you want. However, this is in 'an ideal world' situation which in practice does not exist and most practical dipoles in the real world, while having something of a null off the ends of the wires, have a radiation pattern closer to omnidirectional than the figure-of-eight pattern shown in **Fig 3.2**. In practice this means that it does not really matter too much in which direction the dipole is erected and its orientation can be dictated by the presence of existing suitable supports without worrying too much about the radiation pattern.

All home-made dipoles will probably need to be 'pruned' to size: no matter how carefully measured, 'real life' gets in the way again and unless you are really lucky no antenna is resonant precisely where you expect it to be. Usually they are somewhat too long, which is fortunate, for it is always easier to shorten a wire antenna rather than to have to make it longer by adding bits of wire on. **Table 3.1** shows the approximate 'real life' lengths of half-wave dipoles for SSB frequencies in each of the HF bands. You will probably still need to trim your antenna, even if you use these exact measurements: "your mileage may vary", as they say.

## VARIATIONS ON A THEME

The classic dipole is horizontal, but there is nothing to prevent you mounting a dipole in an inverted-V or sloping configuration (**Fig 3.3**) if that is more convenient in your particular circumstances. A dipole can also be mounted

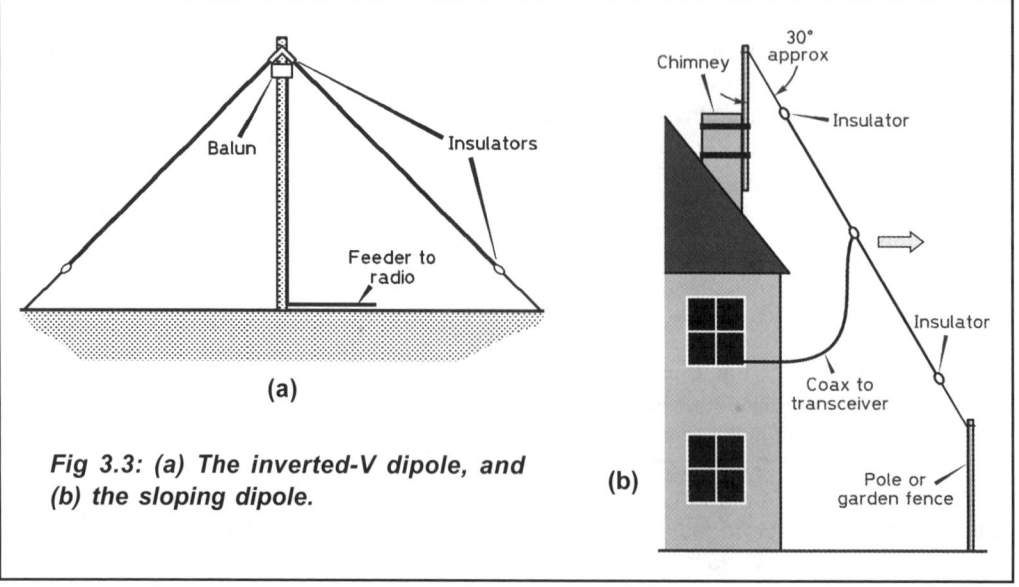

*Fig 3.3: (a) The inverted-V dipole, and (b) the sloping dipole.*

vertically, for example by dropping it from the branch of a high tree. Whatever the configuration, the antenna will still work although the radiation pattern will be different, tending away even more from the horizontal dipole's 'figure of eight' pattern and becoming omnidirectional when the dipole is mounted vertically.

There are some practical and theoretical advantages to these 'variations on a theme': they only require one high support rather than two for the horizontal dipole and, as the antenna becomes more vertically polarised, so the angle of radiation is lowered. However, vertical antennas are inclined to pick up more unwanted local noise than horizontal ones so this advantage may be negated by an increased noise level. The sloping dipole exhibits a small amount of directivity in the direction of the slope, i.e. off the end of the wire, looking from the top towards the low end of the dipole, as shown by the arrow in **Fig 3.3 (b)** – but only a little directivity (this is *not* a beam antenna!)

Due to the height requirement of the support for a sloping or vertical dipole these antennas can only really be considered for the bands from 14MHz and higher in frequency, unless you happen to live in a high-rise apartment block.

We said earlier that a dipole is basically a single-band antenna. While this is true, there are a couple of ways that a dipole can work, or be made to work, on more than one band. The first, and simplest, way is to take advantage of the fact that the low impedance at the centre feedpoint of the antenna occurs not only when it is one half-wavelength long, but also when it is three half-waves long. We can take advantage of this where amateur bands have such a harmonic relationship, i.e. where one band is three times the frequency (and therefore one third of the wavelength) of another, e.g. with the 7MHz and 21MHz bands.

What this means is that a 40m dipole will also work on 15m. Well, sort of. Unfortunately, life isn't quite that simple. The so-called 'end effect' means that a half-wave dipole is actually physically about 5% shorter than the theoretical half-wave length. However, when operating on its third harmonic, the antenna is actually 15% shorter than the physical length of a three half-waves antenna. What this means is that it is resonant higher in frequency than you might expect.

In practice, the way around this problem is to resonate the 40m dipole at the very bottom of the band, 7000kHz, or even make it somewhat longer still, so that the minimum SWR point is actually a little below the bottom of the band on, say, 6980kHz. On 21MHz you will find the minimum SWR point is at the top of the band, around 21450kHz. An external ATU or the internal automatic ATU in your rig should be able to reduce the SWR to 1:1 at your operating frequency of choice in both the 40m and 15m bands. This is one case where an ATU probably *will* be necessary.

Another way of getting a dipole to work on more than one band is simply by connecting two or three dipoles for different frequency bands to the same feeder (see **Fig 3.4**). This arrangement is sometimes called a 'fan-dipole'. In theory any number of dipoles can be connected together in this way, but in practice each one interacts with each of the others in one way or another.

# HF SSB DX BASICS

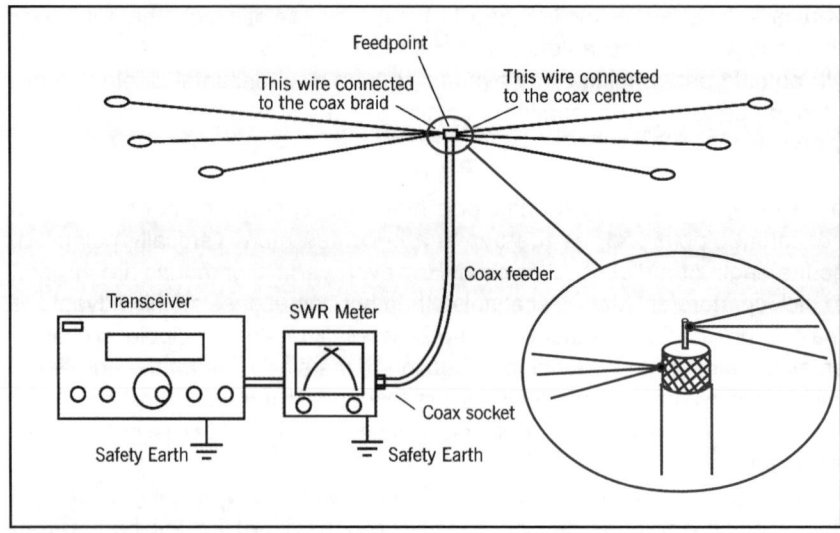

*Fig 3.4: A 'fan-dipole'. Dipoles for two or three different bands are simply connected to the same feeder.*

I have had a lot of success using two dipoles on the same feeder but have found it difficult to get three or more different bands all working together. It is important to separate the dipoles for the different bands physically. For example, if two dipoles are on the same feeder and if you have sufficient space you could have the two dipoles at 90° to each other, or one dipole supported horizontally and the other hanging as an inverted-V beneath it.

The final variation on the theme of a dipole is the *doublet* (**Fig 3.5**). This is simply a dipole fed in the centre with twin feeder (slotted twin, open-wire feeder, ladder line or ribbon) instead of coaxial cable, and matched with an

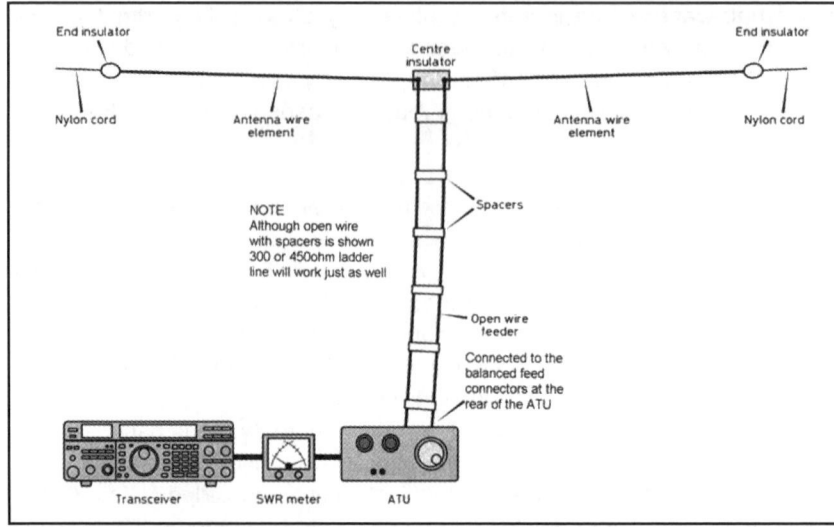

*Fig 3.5: The doublet antenna.*

# 3 — PLANNING YOUR ANTENNA

ATU. The wire can be *any* convenient length but should preferably be around a half-wave long at the lowest intended frequency of operation. While the theoretical impedance of a half-wave dipole is 73Ω (in practice it is likely to be lower than 73Ω and closer to 50Ω) and therefore it matches well to 50Ω coax, the same length of wire will present completely different impedances on other bands, perhaps in the region of thousands of ohms. There is therefore a large mismatch with the 50Ω coax and this is why the dipole is considered to be a single band antenna.

The doublet has two advantages over a coax-fed dipole. Firstly, when used with a wide-range ATU, it becomes a multi-band antenna. Secondly, twin feeder has less loss than coaxial cable, especially at a high level of SWR, so more of the transmitter's power is delivered to the antenna, particularly on the higher-frequency bands such as 12m and 10m. However, there is also the disadvantage that a wide-range ATU *is* necessary; the transceiver's internal ATU is unlikely to work on any band.

## G5RV / ZS6BKW ANTENNAS

The famous G5RV antenna, designed in 1946 by the late Louis Varney, G5RV, is still very popular today, particularly with some beginners who perhaps see it as a convenient way of getting on all bands from 80m to 10m with a single, relatively inconspicuous, wire antenna. It is easy enough to make one, but numerous commercial versions are available from the usual retailers and its wide availability has probably added to its popularity.

The antenna is a type of doublet but of specific length, with a 102ft (31m) horizontal 'flat top' and a vertical matching section preferably 34ft (10.36m) long if made from open wire feeder, or 29ft 6in (8.99m) long if made from 300Ω ribbon feeder. From the end of the matching section, G5RV said that 75Ω twin lead or 80Ω coax can be used to the ATU (these days, 50Ω coax is used instead). Note that a wide-range ATU is required.

G5RV designed the antenna for optimum performance on 20m, where it is a three half-wavelength long dipole. Its multi-band capabilities are something of a bonus, although it should be pointed out that *any* random length of antenna fed with open wire feeder will work on any band with the use of a wide-range ATU, provided it is close to a half-wave (say 0.4λ) in length or longer.

In 2007 Brian Austin, ZS6BKW (now G0GSF), used technology un-available to G5RV – computer modelling – to re-compute the design, while also taking into account the three additional bands at 12, 17 and 30m that were allocated to amateurs at the World Administrative Radio Conference in 1979. He came up with the lengths shown in **Fig 3.6** for an

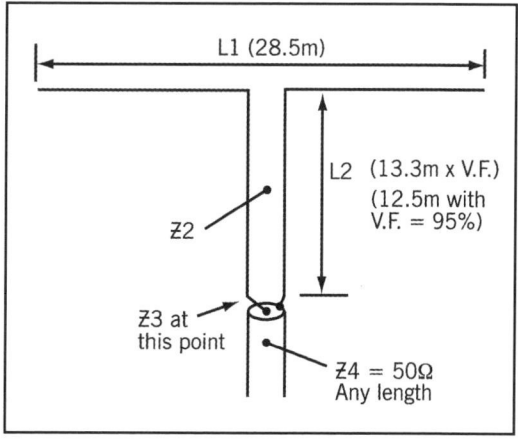

*Fig 3.6: The ZS6BKW antenna.*

antenna that presents a better than 2:1 SWR *without the use of an ATU* in the 40, 20, 17, 12 and 10m bands. It can also be used *with* an ATU on 80, 30 and 15m.

The antenna is 28.5m (93ft 6in) long, and the matching section (or 'series section impedance matching transformer' as G0CSF calls it **[4]**) of 400Ω twin feeder should be 13.3m (43ft 6in) multiplied by the velocity factor of the feeder (so if the velocity factor is 0.95 the length should be 12.6m or 41ft 4in). The dimensions are not overly critical.

If a simple single-band wire antenna is required, it is hard to beat a dipole as high and in the clear as possible. However, if a multi-band wire antenna is really necessary, the ZS6BKW design should be a better bet than the G5RV, as an ATU is not required for operation on five bands.

## VERTICALS

If a dipole is the easiest HF antenna to make yourself, the single-band quarter-wave vertical must run it a close second. It is a little more difficult to get a vertical to work really well, though.

In order to work properly, a quarter-wave vertical must be tuned against some form of ground system. The earth or ground plane forms the 'missing' other half of the antenna if compared with a half-wave vertical dipole. The earth can be thought of as a mirror: if a quarter-wave vertical is placed on top of a mirror the reflection in the mirror is the 'missing' half of a vertical dipole. A vertical antenna tuned against ground in this way is sometimes called a Marconi antenna, after Guglielmo Marconi, the first to utilise such an antenna.

*Guglielmo Marconi (1874 – 1937), who gave his name to the Marconi vertical antenna.*

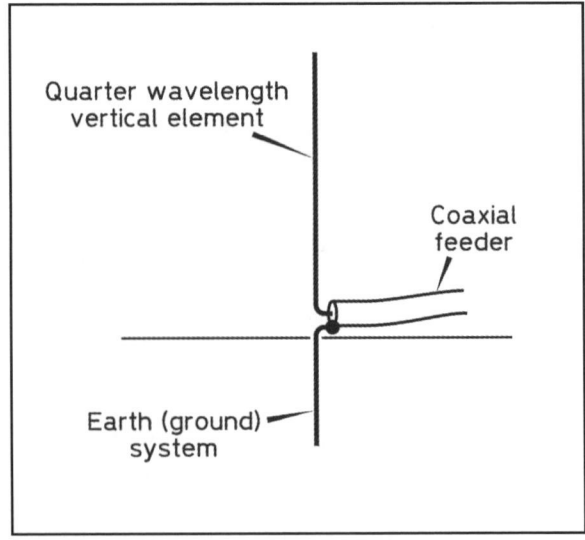

*Fig 3.7: A ground-mounted vertical. Here, the ground system connection is shown as a short earth rod but it is far more effective to use a large number of radial wires as described in the text.*

## 3 — PLANNING YOUR ANTENNA

The quarter-wave vertical can be ground mounted, or elevated, e.g. mounted on a pole or chimney. Let's look at the ground-mounted version first (**Fig 3.7**).

The antenna is a quarter-wave long and is fed by 50Ω coaxial cable at the base. The vertical radiating element is connected to the inner of the coaxial cable and the braid to the earth system, The ground system should be a system of radial wires extending from the base of the vertical radiating element like the spokes of a bicycle wheel, rather than the single ground connection shown in **Fig 3.7**.

The radials are wires which may or may not be insulated and which can be either buried a few centimetres (an inch or two) under the ground, laid on top of the ground, or elevated above the ground. If the radials are on the ground or buried there should be as many radials as possible and preferably they should be as long as possible, though for a particular total length of wire it is better to have many shorter radials than fewer longer ones. The golden rule, though, is 'the more, the merrier' and many amateurs spend a great deal of time and effort laying down more and more of them. There is a law of diminishing returns, though, and many settle for around 32 quarter-wave long radials. The generally accepted maximum is 120 half-wavelength long radials − beyond that there is little or no improvement and the extra work is certainly not worth the effort.

If the radials are elevated it is usual for them to be resonant, i.e. they should also be a quarter-wave long. They should be elevated at least a few feet above the ground. If the radials are elevated, you can get away with using fewer of them: as few as two at 180° to each other, although it is more common to use four, one every 90°. It is even possible to use just one elevated radial (a quarter-wave vertical with a single elevated radial is in fact just an inverted-V dipole rotated on to its side).

If the quarter-wave vertical is itself elevated, e.g. by being mounted on top of a pole, there is no choice *but* to have elevated radials. This is the classic ground-plane antenna (**Fig 3.8**): a quarter-wave vertical radiating element with four quarter-wave long radials which form the ground plane. It is more often used on VHF / UHF (where the radials are short enough to be made rigid and therefore self-supporting) than on HF. On HF the radials can be made of light wire terminated in an insulator and then tied off with string to any four convenient points.

We have already said that, for a simple wire antenna, it is hard to beat a resonant half-wave dipole, whether mounted horizontally, or in an inverted-V or sloping configuration. Why then might one want to use a quarter-wave vertical? The answer is that, for DX working, what is required is a low

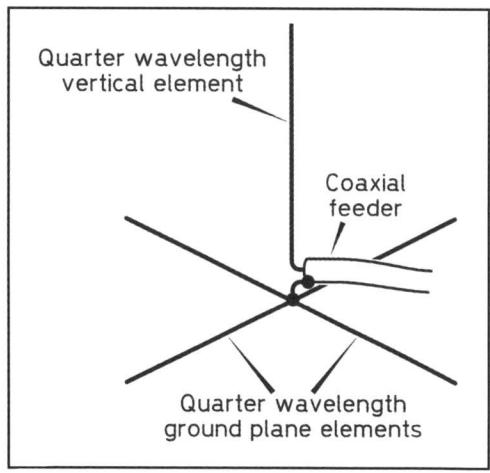

*Fig 3.8: The classic ground-plane vertical: a quarter-wave radiating element with four quarter-wave long radials.*

35

# HF SSB DX BASICS

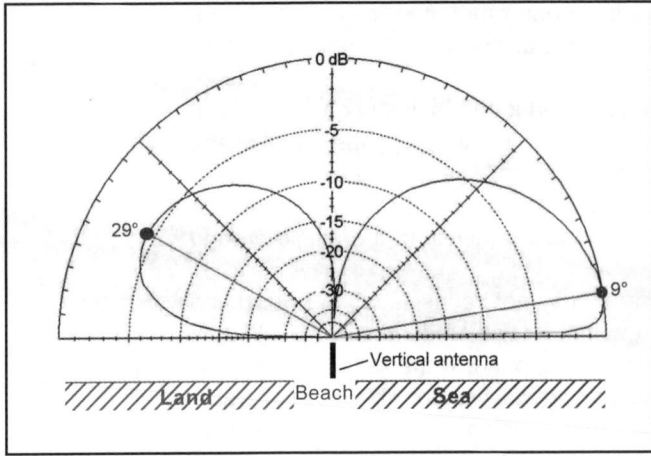

*Fig 3.9: Vertical radiation pattern of a quarter-wave vertical mounted on the edge of the ocean.*

*The perfect DX antenna? A quarter-wave vertical with ground radials mounted within a few metres of the ocean.*

angle of radiation. As discussed earlier in this chapter, a horizontal dipole needs to be mounted high up for this to occur. A vertical antenna, on the other hand, provides low-angle radiation even when mounted at ground level.

However, while a horizontal dipole is inherently an efficient radiator, the conductivity of the earth is of prime importance when it comes to the efficiency of vertical antennas. If the vertical is erected over dry, stony ground it will be inefficient and therefore the size of the ground plane underneath it – the number of radials – assumes great importance. If, on the other hand, the natural earth is of good conductivity, the antenna will work more efficiently. The best natural earth is salt water, and verticals erected very close to the sea therefore perform exceptionally well.

As the ground conductivity increases, so the angle of radiation decreases. What is happening is that instead of the power being applied to the antenna radiating from it in all directions and at all angles from straight up into the sky and down to the horizon, all the power is concentrated into a narrow beam a few degrees above the horizon. This provides a considerable amount of gain at that angle. **Fig 3.9** shows the vertical radiation pattern of a quarter-wave vertical mounted on the edge of the ocean. In this diagram the land is to the left and the sea to the right. The vertical's maximum angle of radiation over the land is at 29° above the horizon but over the sea this angle has decreased to just 9°. That's not all: at 9° above the horizon the radiation is 10dB stronger over the sea than over the land.

This makes the vertical antenna

# 3 — PLANNING YOUR ANTENNA

mounted close to the sea the perfect DX antenna. If, however, you don't happen to live in a beach house, the vertical *can* still be a good DX antenna, but then you will need to put a lot of effort into the radial system.

## THE INVERTED-L ANTENNA

An inverted-L can be thought of as being the 'poor man's vertical'. It is used when it is not possible to put up a full-length quarter-wave vertical, either because the height required is too great for a self-supporting (aluminium or fibreglass) vertical or because there is no support available high enough to suspend a wire from. Since it is relatively simple to make or support full-size quarter-wave verticals for all the bands from 7MHz upwards, the inverted-L is usually only used on the lower-frequency bands such as 80m.

The inverted-L is simply a quarter-wave vertical with the top section bent over horizontally, as shown in **Fig 7.10**. Like the quarter-wave vertical, the inverted-L requires a good ground system of radials to work properly. The vertical section should always be as long (high) as possible.

The horizontal section introduces some horizontal polarisation so, for DX working, the inverted-L is not quite as effective as a full-size quarter-wave vertical. However, for those without the possibility of erecting a full-size quarter-wave vertical (for 80m, for example) it can come a close second.

I once made a dual-band 40m / 80m antenna which worked well on both bands. It started out life as a full-size quarter-wave vertical on 40m, made of lightweight but strong telescoping aluminium tubing which was guyed about

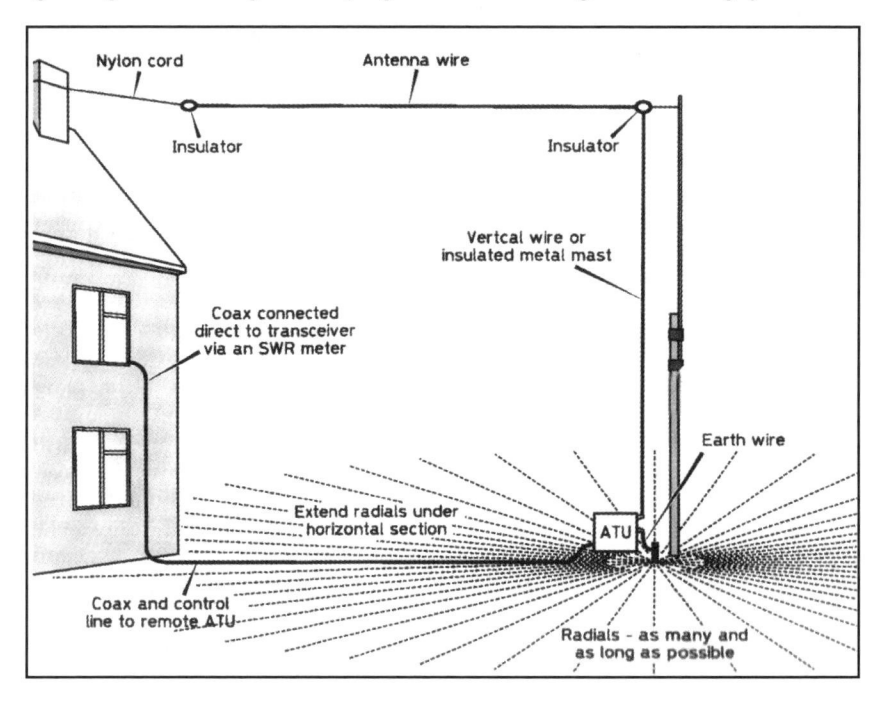

*Fig 3.10: The inverted-L antenna. The ATU makes this a multi-band antenna. Without the ATU the antenna should be a quarter-wave long.*

half-way up. The antenna stood in the middle of the back garden, about 40ft from the back of the house. It was fed at ground level against four buried quarter-wave radials (more would certainly have been better!) The antenna was resonant in the middle of the SSB part of 40m.

To convert it to operation on 80m I connected a very light wire about 8.80m (29ft) long using a small hose clamp at the top of the antenna. At the other end of the wire was a very light-weight plastic insulator and then some garden twine to make up the length required to take it to an upstairs bedroom window at the back of the house. The wire, insulator and twine all had to be very light in weight as the top of the 40m vertical was quite spindly and would not support heavier materials without bending dramatically.

Pulling the wire out horizontally converted the antenna into an 80m inverted-L which resonated around 3790kHz. From my station in south-east England I regularly worked ZL (New Zealand) stations on 80m SSB in the mornings, usually receiving genuine reports of between 55 and 57. This was on the long-path – the long way round the world – a distance in excess of 20,000km. To convert the antenna back to 40m was simply a matter of releasing the twine from the upstairs window and allowing the wire to drop down vertically from the top of the 40m vertical, where it was secured loosely near the base of the vertical.

Undoubtedly a full-size (18.8m / 62ft) vertical would have worked better on 80m SSB, and having more than four 10m (32ft) radials would have worked

## VERTICAL ANTENNA NOTES

Whether commercial or home-made, there are some important points to note about the performance of all vertical antennas:

- The vertical only comes into its own on long-haul DX contacts. If you want to have a big signal around the UK and Europe on 80 and 40m, a horizontal antenna such as a dipole, doublet or G5RV will outperform pretty much any vertical, and sometimes by a large margin . . .
- Quarter-wave verticals only work well if they have a good earth connection in the form of a ground plane of radial wires. An earth spike is not sufficient . . .
- Verticals mounted very close to the sea have a very low angle of radiation and gain at those low angles, so they are great for DX working. Inland, on the other hand, you really need to put the effort into the ground system . . .
- Don't be fooled into believing that because it is a vertical it does not require much horizontal space. Ideally, you need quarter-wave long radials in all directions, i.e. a square 20m x 20m for a 7MHz vertical, with the vertical in the centre of the square . . .
- The performance of a vertical is affected by close objects such as buildings, trees, masts etc (another reason for suggesting that they need at least as much space as a horizontal antenna for the same band) . . .
- Verticals tend to pick up more local noise than horizontal antennas (yet another reason for putting them in their own space). If you live in an electrically noisy neighbourhood, a horizontal antenna may be better – at least for receiving . . .
- Shortened verticals do *not* work as well as full-size quarter-wave verticals . . .
- If you really have very little or no horizontal space for antennas, then consider a half-wave design vertical, but ensure you mount it as high and in the clear as possible.

# 3 — PLANNING YOUR ANTENNA

better too, but for the horizontal space it took up (only 20m or 66ft, including the radials), and its height above ground (a maximum of only 10m or 33ft), I doubt it would be possible to make a better-performing 80m DX antenna.

## COMMERCIAL MULTI-BAND VERTICALS

I started this chapter by urging new licensees *not* to buy a commercial HF antenna at the same time they buy a transceiver, and instead to try building an antenna themselves. However, there is no doubt that commercially-made multi-band HF verticals are very popular, and for good reason too. While it is easy to construct an efficient single-band quarter-wave vertical, it is more difficult to make a successful vertical that operates on multiple HF bands yourself. If you want to have the flexibility of operating on several, or all, of the HF bands but do not have the space for many different antennas, then a commercial vertical may well be the answer.

While there are numerous manufacturers of commercial HF multi-band verticals, in fact there are only two main types: the quarter-wave design and the half-wave design. Both have their advantages and disadvantages.

The quarter-wave type, manufactured by such companies as Butternut, Hy-Gain, Hustler, etc, requires ground radials and is therefore more suitable for ground mounting than for an elevated location such as on top of a pole or a chimney. Quarter-wave verticals *can* be elevated, but the difficulty is in accommodating the radials. Elevated radials need to be a quarter-wave long and, even if there are only two per band, that still means 12 wires of various different lengths for a six-band vertical. In most locations it would be difficult to find suitable places to tie off the ends of 12 different wires around the antenna, an exception being if you have access to a flat roof. Another issue is that some verticals require adjustment, either of coils or lengths of tubing, in order to place the frequency of resonance precisely where you want it. This adjustment can be fiddly and if the antenna is mounted on the chimney for example it could become quite tedious, to say the least, to get it 'spot on'. The quarter-wave design of multi-band vertical is, therefore, more suitable for ground mounting.

The alternative is the half-wave type of vertical, which does not require radials. If you do wish to mount a multi-band HF vertical well above ground, the half-wave design is certainly simpler to erect. The best-known manufacturers of the half-wave design of vertical are Cushcraft and Hy-Gain (Hy-Gain makes both quarter and half-wave designs). In fact these verticals *do* incorporate short spoke-like 'radials' as part of the antenna design, although they are not radials in the usually accepted meaning of a ground plane, but rather a form of capacitance to help in providing a suitable match. Being an electrical half-wave long, and end-fed, the impedance at the base of the antenna is very high and it therefore needs a matching circuit to match the antenna to 50Ω coax. This is provided by the spokes, together with a transformer in the small 'black box' at the base of the antenna. It should be noted that

*Close-up of resonating circuits on a Butternut HF6V vertical.*

although these antennas are electrically a half-wave long, physically they are much less than this and all shortened antennas are lossy to a greater or lesser extent.

The efficiency of a quarter-wave vertical is dependent on its physical length as a proportion of a full-size quarter-wavelength. A full-size quarter-wave vertical will always outperform one that is physically only, say, one-eighth of a wavelength long. Generally on the higher frequency bands this is not an issue, as a quarter-wave vertical on 14MHz is only 5m (16ft) high. It becomes slightly more of an issue at 7MHz, where a full-size quarter-wave vertical is 10m (32ft) high but the length of a vertical is most important on 3.8MHz, where a full-size quarter-wave is 18.8m (62ft) long. Most multi-band verticals covering 3.8MHz are much shorter than this and therefore all are something of a compromise on this band, but one 10m (32ft) long is going to be more efficient than one of only, say, 5m (16ft).

Which work better – quarter-wave or half-wave? I don't know. Probably ground-mounted, side-by-side, the quarter-wave design would be superior *if* it had a good radial system. If a half-wave design vertical is mounted high up and in the clear and a multi-band quarter-wave design vertical is ground mounted it might be a more close-run race.

## LOOPS

Having discussed half-wave dipoles and quarter-wave verticals it is time to look briefly at full-wave loops. Loops can be constructed in the horizontal plane but for DX work they are more usually made in the vertical plane as this can (depending on how the loop is fed) give a low angle of radiation. The full-wave loop also has a small amount of gain over a dipole.

The loop can be more or less any shape: the ideal shape is a circle but as this is mechanically difficult to construct most loops are either rectangular (preferably square or nearly square) or triangular in shape. A square (or a diamond-shaped) loop is usually called a quad loop (**Fig 3.11**) whereas a triangular one is a delta loop (**Fig 3.12**).

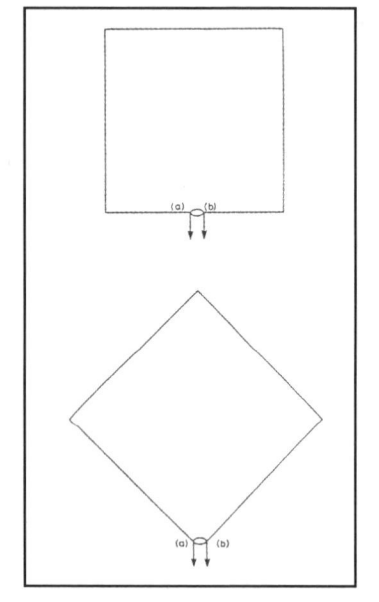

*Fig 3.11: Square and diamond-shaped quad loops.*

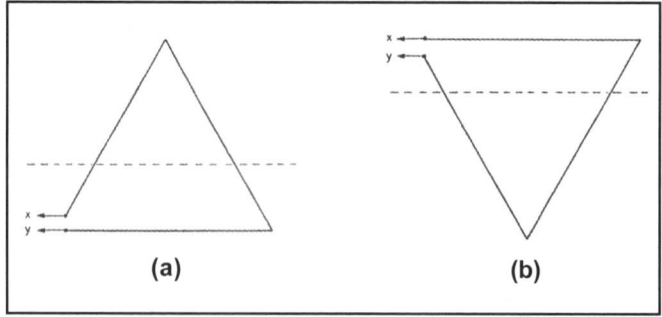

*Fig 3.12: Delta loops, (a) 'apex up' and (b) 'base up'. The dotted lines show the effective average height above ground for each antenna.*

## 3 — PLANNING YOUR ANTENNA

For home-made wire antennas, the delta loop is probably more popular than the square or diamond-shaped quad loop. One reason for this is that only one high support is required for the 'apex up' configuration as shown in **Fig 3.12 (a)**. To provide the low angle of radiation suitable for DX working, the delta loop needs to be fed at one corner, as shown in **Fig 3.12 (a)** and **(b)**, or close to the corner but a short way along one of the sloping sides. (Feeding it in the centre of the horizontal wire would provide high-angle horizontal radiation.)

The formula for calculating the length of the loop is normally given as $1005/f$ (MHz) in feet or $306/f$ (MHz) in metres, e.g. $1005/14.2 = $ 70ft 9in, or $306/14.2 = 21.55$m for a delta loop resonant in the SSB part of 20m, at 14200kHz. These formulas normally produce a loop that is somewhat too long but, as with the dipole, it is always easier to shorten the antenna than to have to add more wire later.

## BEAM ANTENNAS

So far we have looked only at single-element wire antennas and verticals made of wire or thin tubing. But, if you have the possibility to put up some sort of beam antenna, there is no doubt that, for DX working in particular, it will give you a great advantage over a single-element antenna.

A beam antenna is one with both *directivity* and *gain*. See **Fig 3.13**, which compares the theoretical horizontal radiation pattern of a dipole with that of a beam antenna, in this case a 4-element Yagi. It can be seen that the major lobe of the Yagi is much greater than that of the bi-directional dipole, with little radiation off the back and the sides of the antenna.

A Yagi beam consists of at least two *elements*: a *driven element* which is usually a half-wave dipole, and at least one *parasitic* element. In a 2-element Yagi the parasitic element is usually a *reflector*. A 3-element Yagi will also have a *director*. If a Yagi has more than three elements there will usually be multiple directors: it is normal for there to be just one driven element and one reflector, no matter how many elements there are in total.

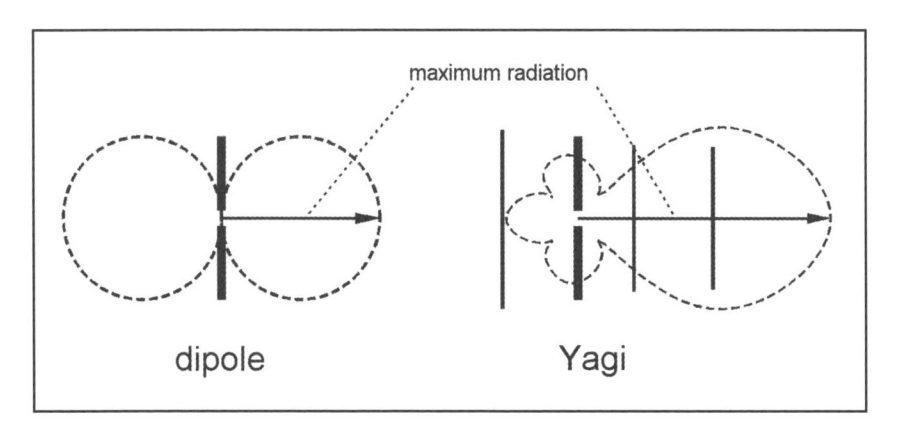

*Fig 3.13: Comparison between the horizontal radiation patterns of a dipole and a Yagi beam antenna.*

> ## GAIN FIGURES DEBUNKED?
>
> The gain of a beam antenna is expressed in decibels (dB), but it must always be given with reference to something, otherwise the figure is quite meaningless.
>
> Often the gain is quoted with reference to a dipole in free space, e.g. a particular antenna might be said to have 5.0dBd gain. But it can also be given with reference to an *isotropic radiator*, a theoretical point source that radiates at the same intensity in all directions. That same antenna might be said to have 7.15dBi gain. A (theoretically loss-less) dipole in free space has 2.15dBi gain, so it is easy to convert between gain figures quoted in dBd and dBi simply by adding or subtracting 2.15dB from the quoted figure.
>
> Real life is a little more complicated though, because any antenna, when placed over the real earth, actually exhibits *more* than the theoretical amount of gain, due to so-called 'ground gain'. The amount of this extra ground gain varies depending on the characteristics of the ground and the height of the antenna above ground, so it is not easy to quantify, but it *can* be as high as 8dBi for a dipole, i.e. an additional 5.85dB.
>
> Antenna gain figures should therefore be quoted in units of either dBd or dBi, but if a gain figure is quoted as, say, '9dBd over typical ground' you have no idea what this really means *unless* the amount of additional ground gain is also given. '9dBd gain including typical ground gain of 4dB', simply means the true gain is 5dBd (or 7.15dBi).

The gain of a Yagi is not primarily dependent on the number of elements but rather on the boom length of the antenna, i.e. the physical spacing between the elements.

A *cubical quad* is a beam antenna made from a full-wavelength long quad loop in either a square or diamond-shaped configuration plus a parasitic reflector element, also a loop of the same shape and physically or electrically approximately 5% larger. This is a 2-element quad: additional director elements may be added to make 3-element or greater quads.

Since a single-element quad loop has a small amount of gain over a dipole, a 2-element cubical quad has about the same gain as a 3-element Yagi. In other words, the same amount of gain is achieved in a quad with a boom only about half as long as a Yagi's; but the disadvantage is that a quad has height as well as length and breadth. It can therefore be a more unwieldy structure than a Yagi. Because a quad increases in size in all three dimensions as you go lower in frequency, a 20m quad – even a 2-element one – is a big antenna. The reverse is also true, though, making a quad for 10m or 12m a fairly compact antenna.

While it is perfectly possible to build both Yagis and quads, the size and physical construction make home construction of any HF beam a fairly major project. Designs for traditional aluminium tube Yagis can be found in *The ARRL Antenna Book* [1] but in recent years there have been several

# 3 — PLANNING YOUR ANTENNA

interesting designs published for wire beams, including the Spiderbeam **[5]**, the Moxon Rectangle **[6]** and the Hexbeam **[7]**. All three designs use a frame made of fibreglass poles which act as spreaders to keep the wire elements at the correct spacing with respect to each other. The fibreglass and wire construction makes these designs much lighter than traditional all-metal Yagis and all three would be good projects for those who wish to take up the challenge of making their own HF beam antenna.

## COMMERCIAL BEAMS

Most amateurs, if they plan to use a beam antenna at all – and whether they be newcomers or the very experienced – opt to purchase a commercial beam.

Many British radio amateurs though, living in typical suburban houses with small gardens, do not even consider the possibility of putting up an HF beam antenna due to its size and the perceived objections they will face from their neighbours. However, when considering an HF beam almost everyone has either a 3-element triband aluminium Yagi or a 10m to 20m 2-element cubical quad in mind. It is true that these *are* big antennas and their size does put off many. There *are* smaller commercial beams available, though, which should definitely be considered.

The first worth considering is a monoband 3-element Yagi for 10m or 12m. With typical dimensions of only 5.5m wide by 2.45m long (and weighing only 4 or 5kg) a 10m beam is not really a big antenna at all (and, if mounted at roof height or above it will appear even smaller!) A 12m Yagi is only slightly larger. It is also possible to buy cheap Yagis intended for 11m CB and modify them for use on 10m (e.g. **[8]**).

Unfortunately now (in late 2015) we are on the downward slope of Solar Cycle 24, so propagation is likely to be poor, and deteriorating, on the 10m and 12m bands for the next few years. Most DXers will therefore want a beam antenna that covers 20, 17 and / or 15m and yet is still not too big. The Moxon Rectangle **[5]** is a good design, but there are few commercial versions available for HF and the design does not lend itself well to multi-banding.

One design that *does* is the Hexbeam **[7]** and a number of commercial manufacturers around the world (e.g. **[9]**) offer either 5-band (20 – 10m) or 6-band (20 – 6m) versions. The Hexbeam is a 2-element wire Yagi on each band, the unusual design of which – like the ribs of an upturned umbrella – allows for optimum spacing between the elements on each of the bands. The original design was improved upon by Steve Hunt, G3TXQ, who came up with a

*A portable version of the 6-band Hexbeam.*

broadband version of the beam. The two elements are bent into 'C' and 'W' shapes, making an antenna that is only 6.5m across and with a turning radius of 3.25m.

The fibreglass spreaders and wire construction allow for a very lightweight antenna. One version weighs in at only 6kg, making it possible to install the Hexbeam on a lightweight push-up mast and rotate it with a cheap "TV-type' rotator.

## REFERENCES

[1] *The ARRL Antenna Book*, 22nd edition, ARRL, available from www.rsgbshop.org

[2] *RSGB Antenna File*, available from www.rsgbshop.org

[3] *Successful Wire Antennas*, edited by Ian Poole, G3YWX, and Steve Telenius-Lowe, 9M6DXX (now PJ4DX), available from www.rsgbshop.org

[4] 'Technical Topics', *RadCom* (RSGB), May 2007.

[5] DF4SA's Spiderbeam site: www.spiderbeam.com

[6] KD6WD's Moxon Rectangle site: www.moxonantennaproject.com

[7] G3TXQ's Hexbeam site: www.karinya.net/g3txq/hexbeam

[8] Sirio SY 27-3 and SY 27-4 CB antennas: www.nevadaradio.co.uk/cb/cb-antennas/cb-base-antennas

[9] Hexbeam UK (G3TXQ-design Hexbeams manufactured by MW0JZE): www.g3txq-hexbeam.com

# 4 The SSB DXer's Transceiver

IN THIS CHAPTER we look at those criteria that are of interest to all types of operator. The next chapter includes some features that are often overlooked but which are of specific importance to the SSB DXer.

If you were starting from scratch, which transceiver would you buy? There is no 'best' transceiver but, while ostensibly similar (they all transmit and receive SSB and other modes on at least 10 to 80m), many of them do have differing features or individual characteristics. Some of these features may be of no importance at all to you, but they may well be to a different type of operator. A good example of this is the built-in second receiver that some high-end transceivers have included as standard. A DXer might consider this a very useful or even an essential feature, whereas an operator who never chased DXpeditions and only ever had 'ragchews' or called in to 'natter nets' might find this an expensive and completely unnecessary distraction.

All current 'HF' transceivers cover all the HF bands plus 160m and most also cover 6m. Some also include 2m and 70cm and at least one covers the 4m band. Similarly, other than a few specialist and usually QRP(low power) transceivers, all are multi-mode, in other words they will transmit and receive CW and AM (and in most cases FM and perhaps data modes) as well as SSB. Your choice of transceiver will partly be dependent on whether the VHF and UHF bands are important to you or not, and to some extent whether you plan to use other modes as well as SSB. It's a balancing act: a rig that is good on SSB may only offer mediocre performance on CW, for example. If you intend to operate only SSB, or will only make the occasional excursion on to CW, this will not matter too much.

While some criteria (such as receive performance) are equally important to all operators, others (such as whether or not a transceiver has full break-in – 'QSK' – on CW) are of no interest at all to the SSB operator. Likewise, while it might be important to an operator who also wanted VHF / UHF capabilities, for the purely HF operator it is not really relevant whether or not a particular transceiver has 2m and 70cm included.

## EQUIPMENT CRITERIA

There are quite a few important criteria to look at when considering the purchase of any transceiver. In no particular order these include:

- Receive performance
- SDR or 'traditional'?
- Base, mobile, or portable?
- Power output
- Internal ATU
- Front panel display
- 'Bells and whistles'
- Price
- New or second-hand?
- Noise blanker
- DSP / Digital noise reduction (DNR)
- Dual or twin receivers
- SSB speech quality
- Transmission audio tailoring
- Speech processor
- Digital voice keyer
- Monitor facility

Many of these criteria are equally relevant to all operators, while some may be of little or no interest to those who do not intend to be DXers, but prefer to chat across town on 80m, or perhaps across Europe on 20m. Therefore, since this is *HF SSB DX Basics*, the comments in this chapter are made from the perspective of an SSB DX operator.

## RECEIVE PERFORMANCE

For many serious HF DXers a transceiver's receive performance is the single most important consideration when deciding which piece of equipment to buy. Within this single category, there are numerous parameters to consider, including sensitivity, selectivity, dynamic range, image rejection, etc. This is not the place to go into any great detail on the meaning of these terms but, for the moment, suffice to say that what they are measuring is the ability of a receiver to single out a particular wanted weak signal among numerous stronger unwanted signals and to make it readable.

In brief, sensitivity can largely be ignored, because at HF it is not really an issue: just about all commercially-made transceivers have adequate sensitivity these days.

Selectivity, the ability of a receiver to home in on the signal required and reject other nearby signals, is determined by the receiver's filters, which could be crystal or mechanical or, these days, is more likely to be implemented digitally using DSP (digital signal processing). If the transceiver has IF DSP filtering there is likely to be a wide range of IF selectivity bandwidths available to the operator.

Dynamic range is the difference between the weakest and the strongest signal that the receiver can cope with and is measured in decibels (dB). The higher the figure, the better: anything above 100dB is excellent.

Image rejection is the ability of the receiver to reject the image signals that are generated within the receiver during the mixing process. If image signals are not suppressed they will appear as interference on the wanted signal. Once again, this parameter is measured in dB and the higher the figure the better.

However, even if you have a good knowledge of the meaning of these measurements and – just as importantly – how to interpret them, the best way to determine how well a receiver performs is to compare two pieces of equipment side-by-side on the operating desk and using the same antenna. They may well sound *different*, but the question really is whether one can receive readable signals that the other one cannot, and probably 98% of the time you will notice little or no difference in this respect between two different models of transceiver. The remaining 2% of the time is when the one with the better receiver will be able to winkle out the weakest of signals buried in interference. Such top performance often comes at a price though, and, even if you have aspirations to be a top DXer you will need to ask yourself whether paying, say, twice the price is worth it for that 2% of the time.

The fact is that these days all commercial transceivers have pretty good receiver performance, although it is true that some are better than others. If you know how to interpret them correctly, you can get a good idea of how well a receiver will perform by looking at measurements of the various parameters, preferably those made by an independent reviewer such as Peter Hart, G3SJX [1], whose excellent and detailed reviews of rigs have been published regularly in *RadCom* for over 35 years now.

## SDR OR 'TRADITIONAL'?

Before we look at some of the other important criteria to be considered when choosing a transceiver for HF SSB DX work, a word or two about Software Defined Radio, SDR. The term can be used of both receivers and transceivers and it refers to the basic design of the equipment: with SDR equipment computer software, rather than discrete components, is used to provide the filtering, modulation, demodulation etc.

In amateur usage, however, the term SDR has led to some confusion. When referring to SDR transceivers or receivers many amateurs mean only equipment such as the FlexRadio series of transceivers that also require a high-end computer in order to provide the processing power. Such equipment

*The FlexRadio Flex-6700 SDR transceiver.*

# HF SSB DX BASICS

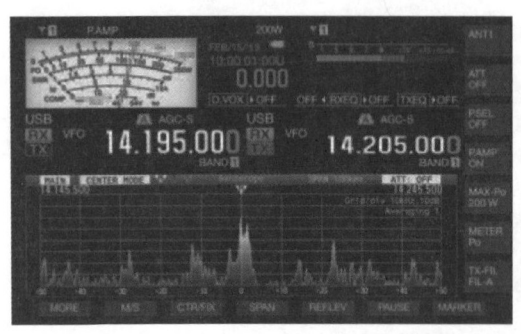

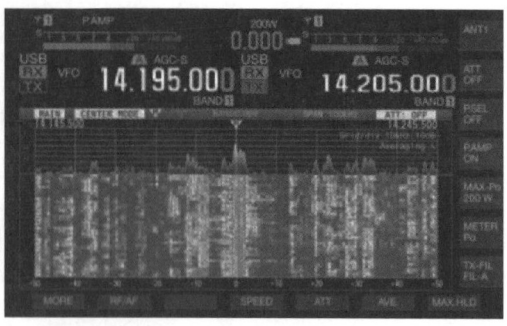

*Kenwood's top-of-the-range TS-990S transceiver features a front panel which can be configured to show a bandscope (left) or a waterfall display (right).*

is, often literally, a 'black box' with no external controls, simply a connection to the computer and sockets for a microphone, Morse key, headphones and antenna. The SDR black box has no front panel, instead the computer screen becomes a 'virtual front panel' and the transceiver is controlled by software, using the PC's mouse and keyboard.

Where the confusion has arisen is that some of the new designs of apparently 'conventional' transceiver *also* employ SDR technology, but using an 'embedded system' (with the processor built into them), so they do not require a separate stand-alone computer in order to be able to function (e.g. the Kenwood TS-990S: see photos above). Such equipment has a conventional looking front panel, with a frequency display, tuning knob, AF and RF gain controls etc but, inside, the technology is still SDR.

One of the great advantages of SDR is that it allows spectrum either side of the frequency to which the equipment is tuned to be displayed on a screen as a *bandscope* or, alternatively, as a *waterfall* display. While these displays undoubtedly look best on a large computer monitor, all the major manufacturers are now producing equipment in which a bandscope and / or waterfall display can be shown on the front panel of the transceiver.

Only time will tell whether the 'black box' design of SDR equipment, with its requirement for a separate computer, will take over from the more traditional type of transceiver with their large tuning knob(s), manually controlled AF and RF gain controls and push-button band switches. Some amateurs have already decided that they *prefer* to control their transceiver using a computer. However, my own belief is that the great majority of radio amateurs will continue to prefer the convenience of a set of controls physical designed specifically for the job in hand.

## BASE, MOBILE OR PORTABLE?

These days, HF transceivers tend to come in three flavours: 'Base station' or 'Mobile', with a few rigs which fall somewhere in between that might best be called 'Portable'.

What are the differences between these three types of transceiver? Firstly and most obviously, base station transceivers are bigger and heavier than their mobile siblings. The trend in recent years is for the top-performing (and

## 4 — THE SSB DXER'S TRANSCEIVER

...transceivers to be large, table-top affairs ...kg and sometimes a lot more. (Many of them ...n as much as they have a carrying handle on the ...ot want to have to carry them too far!) Most base ...also have a built-in mains power supply unit (which can ...the rig's overall size and weight) and therefore they are ...: they operate from 220 – 240V AC mains. (The North ...sions work from 110 – 130V AC, or they may have a dual ...wer supply with a switch so that they can be operated in countries ...either mains voltage is the standard.)

'Mobile' sets are just that: they are intended to be operated from cars, so they are designed to work from a car battery of 12V – 15V DC (the standard operating voltage is 13.8V). There's nothing to stop you using a mobile rig from home, but you will need an external mains power supply unit capable of providing 13.8V DC and with a current rating of up to about 22 or 25A on peaks.

In stark contrast to the base station rigs, the trend for mobile transceivers is for each generation to be smaller and lighter than the last. As a result of this, base stations are usually far superior ergonomically to mobile sets. The latter may have front panels so small that only a limited number of controls can be fitted on them. The manufacturers' way round this dilemma is to have all but the most frequently used controls available only via a 'menu' system. What this means in practice is that a function such as, say, reducing the power from 100W to 5W in order to make a QRP contact

*Size comparison between base and mobile transceivers: the Yaesu FT-2000 (bottom) and the Yaesu FT-857D (top), with microphone. The FT-2000 has a built-in mains power supply unit and ATU; the FT-857D has neither. Both provide 100W output but, ironically, it is the tiny FT-857D that contains two additional bands, 2m and 70cm, in addition to all the bands from 160 to 6m that are provided by the FT-2000.*

requires one or two button presses to select
turning a knob to get to the correct menu nu
knob in order to reduce the power level, then p
the change that has been made and to return to
contrast, on the base station transceiver there will
dedicated 'Power' knob which can be adjusted up and do
is taking place. Many other functions, such as changing th
setting, adjusting the transmitter's speech processor, or ope
the two VFOs, may require a similar number of button presses
adjustments in mobile sets. It is not unusual to find that operators
cannot remember what is required to achieve some of the lesser-us
functions and therefore they need if not the full operating manual then at
least an *aide mémoire* available by the rig at all times.

Many base station transceivers have a built-in automatic ATU, most mobile transceivers do not. An internal ATU can be a useful feature but the lack of one is perhaps less important than it might at first appear, as we shall discuss later.

Mobile transceivers are designed to be operated with mobile antennas, such as electrically-short vertical 'whips'. To compensate for the relative inefficiency of the antenna, the rig is designed with high sensitivity. This may sound all very well in theory, but if used at home on a high-gain antenna (such as a quad or Yagi beam on the higher-frequency bands, or a full-size dipole up high and in the clear on 40 or 80m) some transceivers may be subject to overloading, with spurious ('phantom') signals appearing in the receiver and causing interference to the real signals that you want to hear. This is not necessarily as bad as it sounds, because almost all transceivers these days have a built-in front-end attenuator which can be switched in when required. This decreases the level of the signal going into the transceiver from the antenna and thus eliminates, or at least reduces, the overloading.

Generally speaking the overall receive performance of base stations is superior to that of mobile sets. Like mobile units, 'portable' transceivers do not usually have a built-in power supply unit but they are generally physically larger than mobile transceivers. They are therefore designed to be normally operated from the home station, where there is likely to be a suitable power supply already *in situ*. However, as they are smaller and lighter they are more suitable than base station transceivers for operating on field days, from caravans or on DXpeditions.

## POWER OUTPUT

Most HF transceivers, whether they be base station, portable or mobile models, have a power output of 100W. There are a few exceptions, though. There are low-power transceivers about, such as the Yaesu FT-817 (5W out), Icom IC-703 (10W) and the Elecraft KX3 (also 10W).

Some operators get a kick out of operating with low power ('QRP' operation, usually defined as up to 5 watts output) and the GQRP Club **[2]** has a great following. However, most QRP operators tend to stick to CW and data modes such as PSK31, as these are much more effective than

## 4 — THE SSB DXER'S TRANSCEIVER

SSB at the low signal levels often associated with low power operation. Rigs such as the FT-817, IC-703 and KX3 are ideal for the dedicated QRP operator, but unless you *never* wish to operate with higher power than this, it is not really worth considering purchasing a transceiver with such a low output power. Even if you are at present a Foundation licensee and limited to a maximum of 10W output, you should have aspirations to upgrade your licence, first to Intermediate level and then to a Full licence. If you are a DXer, once you upgrade you will almost certainly want to use higher power than 10W, so do take this into consideration when choosing your transceiver.

This is particularly the case for SSB DXers: using only 5 or 10W, DXing on SSB is *hard* work, far more so than on CW or PSK for example, unless you are lucky enough to have some spectacularly good antennas.

At the other end of the scale, there are a few transceivers that provide 200W output and at least one that can give 400W out. Is this sort of power level necessary? The answer will depend on your style of operating. If you were to only have 'ragchews' with stations around the UK on 80 or 40m, or around Europe on 40 or 20m, then the answer is probably "no": 100W will be sufficient. But for the DXer (or contester) there will certainly be times when it would be nice to have more than 100W available.

Going from 100W to 200W, i.e. doubling the power output, equates to an increase in signal level of 3dB. An increase of 2dB is only *just* noticeable at the far end, so on the face of it there seems little point in going for a 200W transceiver over a 100W transceiver, all other things being equal. Going from 100W to 400W, on the other hand – quadrupling the power output – provides an increase in signal level of 6dB. This is equivalent to 1 or 2 S-units, depending on how the receiver's S-meter is calibrated (traditionally an S-unit was taken to be 6dB, but most modern transceivers have only 3dB per S-unit). This sort of increase *is* worth having. It may not appear to be important if signal levels are around S9 at the 100W level, but if the signal is only about S4, an increase to S6 might make all the difference between being heard and being buried in interference.

The most common way of increasing one's power output above 100W is to use a separate linear amplifier after the transceiver. Most commercial

*The Icom IC-7700, one of the new breed of base station transceivers to provide a 200W output level.*

51

linear amplifiers are capable of power outputs of between 500W and 1500W, depending on the model and the design. In order to achieve that power output, they need to be 'driven' (by the transceiver) with an input power of anywhere between about 60W and 100W (again, depending on the design). A 200W output transceiver, therefore, is unnecessary if you have a linear amplifier: 100W is ample power to drive the linear amplifier to its full power output. A 400W transceiver would be even more 'overkill' – if you have a linear amplifier, that is.

In the UK the maximum power limit is 400W, so does it not make sense to cut out the linear amplifier altogether and instead buy a transceiver with 400W output? There are a few reasons why this may not be the best option, logical though it sounds. Firstly, a 400W-output transceiver might only provide a maximum of 400W output when there is a perfect 1:1 match between the transceiver and the antenna system. If the SWR is above 1.5:1 or 2:1 the power output might well be reduced. This can be counteracted by using the transceiver's internal ATU, but this also introduces a loss, sometimes of 10 or even 20%. Secondly, the Full UK licence allows 400W input at the antenna, i.e. you can use more than 400W out of the transmitter if there is loss between the transmitter and the antenna itself. Obviously there will always be *some* loss: the question is 'how much?' While it is always good practice to ensure that feeder losses are kept to a minimum, with a long feeder run it is possible that on 28MHz there may be a loss of, say, 3dB between the final amplifier and the antenna, when insertion loss of an ATU and a low-pass or band-pass filter (see **Fig 4.1**) are included. With 3dB loss it would be permissible to run 800W out of the amplifier so as to provide 400W at the input of the antenna. It is therefore important to be able to measure the power at the antenna, or know precisely how much loss there

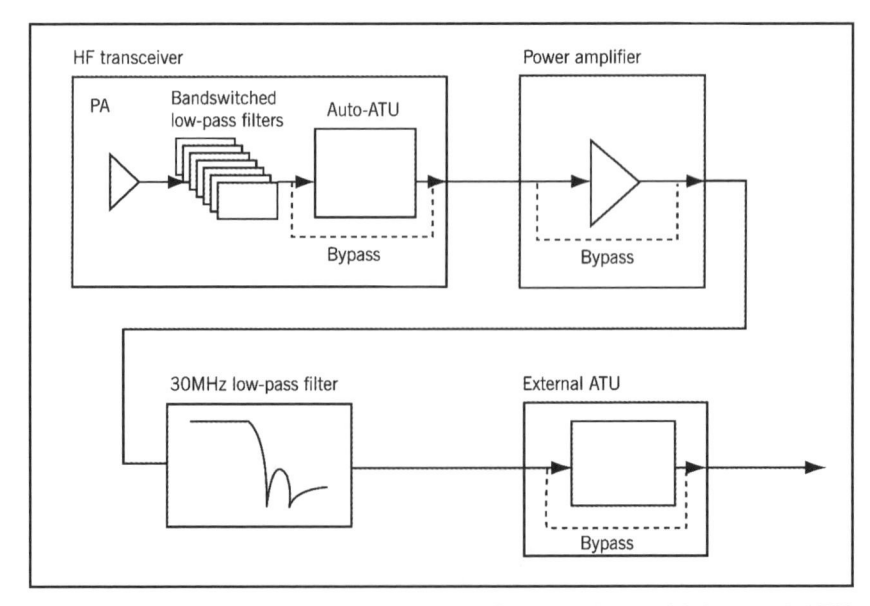

*Fig 4.1: Typical HF station consisting of transceiver with internal ATU, external power (linear) amplifier, low-pass filter and external ATU.*

# 4 — THE SSB DXER'S TRANSCEIVER

is, so as not to exceed the power limit. Thirdly, choice: there is a very limited choice of transceivers available that give 400W out. Finally, cost: a 400W-output transceiver is likely to be very much a top-of-the-range rig and cost more than a modest, though perfectly acceptable, 100W transceiver and medium-sized linear amplifier capable of the 500 – 800W necessary if you really do wish to run 400W at the input of the antenna.

*The Kenwood TS-480HX, the smallest and, at 3.7kg, the lightest, 200-watt output HF transceiver currently available.*

All the above suggests that the best option is a 100W transceiver: many operators do not require more power than this, and those who do are usually best served by using a separate linear amplifier, which only needs 100W (or less) of drive power.

However, there *are* a couple of reasons for buying a higher-power transceiver. If you intend to do portable operation or go on DXpeditions, for several reasons it may not be possible to take a linear amplifier with you – for a start they are generally large and very heavy. Under those circumstances, a 200W-output transceiver will give your signal just that little extra edge over a 100W station. Unfortunately for the lightweight traveller, most of the 200W radios are also pretty large and quite heavy. There is one exception, though: the Kenwood TS-480HX weighs only 3.7kg, although it should be noted that due to the higher power output it requires a 40 – 45A 13.8V power supply (or two identical 20 – 23A PSUs). The other reason, once again, comes down to cost. If you want all the power that you can muster yet have decided that you cannot justify the expense of a separate linear amplifier, a 200W transceiver might be the answer. Certainly some 200W transceivers are cheaper than some 100W rigs that have a higher specification in aspects other than power output.

## INTERNAL ATU

ATU stands for antenna (or aerial) tuning unit or, more correctly perhaps, 'antenna-system tuning unit'. It is also sometimes known as an 'antenna matching unit'. Some transceivers have a built-in automatic ATU, while others do not. In some, it is an option available at extra cost. While it can only really be an advantage for a rig to have an internal ATU, in my opinion it is not a *great* advantage and should not be seen as a necessity, nor even a particularly important feature.

There are several reasons for this. Firstly, many amateurs only ever use antennas that have an impedance of close to 50Ω, in which case an ATU is unnecessary. Secondly, transceivers' internal automatic ATUs can generally only match SWRs of up to about 3:1, i.e. impedances of approximately 16 to 150Ω. Admittedly a few can do much better than this, but rarely can they cope with the very wide range of impedances that might be provided when trying to match, for example, a long-wire antenna or multi-band doublet. For that you will need an outboard (external) wide-range ATU (either manual or automatic). Thirdly, putting an internal ATU into circuit can introduce a power loss of up to 10 or 20%.

Finally, if you ever intend to use a linear amplifier, the linear goes between the transceiver and the antenna, so the transceiver's internal ATU will be of no use in matching to the antenna system: what the ATU will 'see' is the input of the linear amplifier, which should itself always be close to 50Ω. If a linear amplifier is in use what is needed then is an automatic ATU at the output of the linear amplifier, or an outboard ATU between the linear amplifier and the antenna system, as shown in **Fig 4.1** (however, once again, if the antenna has an impedance close to 50Ω the ATU is unnecessary anyway).

Where an internal ATU *is* of use is as a 'line-flattener'. Transceivers need an SWR of better than 2:1 (or sometimes as little as 1.5:1) in order to deliver their full power to the antenna. If you are using an antenna with a high 'Q' (one with a sharp or steep SWR curve), the SWR may rise to above 1.5:1 or 2:1 in part of the band, even if it is an ideal 1.0:1 at another part of the band. Where the SWR rises, it is likely the transceiver will automatically reduce its power output. The internal ATU can be useful to 'flatten' the SWR curve and provide a 1:1 SWR match, thus enabling the transceiver to put out its maximum power on a wider range of frequencies than would otherwise be the case. In practice, these circumstances are most likely on 80m where, if the antenna is resonant at, say, 3800kHz, the SWR may be above 1.5:1 or even 2:1 at, say, 3650kHz. Here, the internal ATU could be brought into service to reduce the SWR seen at the transceiver to 1:1 at 3650kHz.

Note again, though, that an internal ATU in the transceiver is of no use at all if you are using a linear amplifier.

## FRONT PANEL DISPLAY

At one time, the front panel display on a transceiver was an analogue dial from which you could read the operating frequency to within 1 or 2kHz if you were lucky. In the 1970s along came digital frequency readout. Frequency synthesisers came in at about the same time and their implementation meant that you could be reasonably certain that the digital frequency readout was accurate to within about 0.1kHz or so.

Transceivers' front panels continued to develop over the years and now some, such as Yaesu's popular FT-450, have a simple block diagram of the receiver front end as part of the front panel display (see **Fig 4.2**). This useful feature gives a clear indication of the signal path and the current operating parameters of the receiver at a single glance.

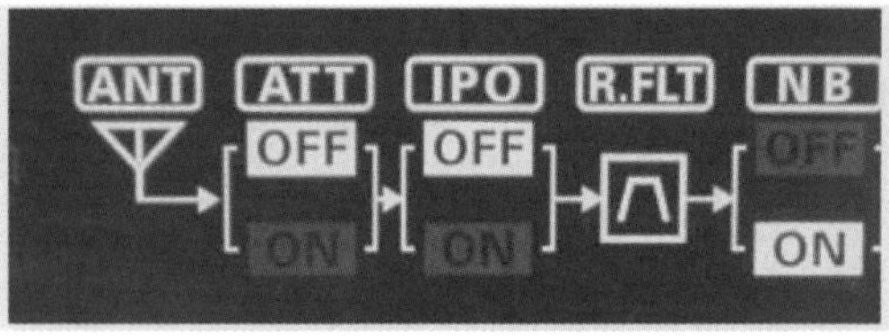

Today there may be several dozen

*Fig 4.2: Yaesu FT-450 front panel signal path display. In this case the signal goes from the antenna to the roofing filter, with the attenuator (ATT) and intercept point optimisation (IPO) both switched off, to the noise blanker (NB), which is switched on.*

# 4 — THE SSB DXER'S TRANSCEIVER

operating parameters displayed on the front panel, including VFO A operating frequency, VFO B operating frequency, mode, pre-amplifier 1 on/off, pre-amplifier 2 on/off, amount of RF attenuation, power output, AGC setting, RIT offset, S-meter, noise blanker on/off, filter bandwidth, roofing filter and much more besides.

*The Icom IC-756Pro, introduced in 2000, was the first to use a colour TFT front panel display.*

The 'big three' Japanese manufacturers (Icom, Yaesu and Kenwood) now all use colour thin film transistor (TFT) liquid crystal display screens on at least some of their transceivers. This allows the front panel to be used to display, for example, a bandscope. The technology is no longer new: it was pioneered by Icom with the introduction of their IC-756Pro transceiver as long ago as 2000. (However, the respected US manufacturer Elecraft continues to use a traditional monochrome LCD screen.)

There's no doubt that colour TFT screen displays are useful and attractive. Just one more factor to consider when deciding which transceiver to buy!

## 'BELLS AND WHISTLES'

'Bells and whistles' is the rather derogatory term for the features that come with your transceiver that you may not really need. However, one man's bells and whistles are another man's essential requirements, so don't ignore these features entirely: you may find some that fit in well with your particular operating style. I have already said that a transceiver's internal automatic ATU is, to me, not a particularly useful feature whereas (in my opinion) an effective noise blanker and dual receivers (discussed later), as well as a good speech processor and the Monitor facility (see Chapter 5) are all very useful, if not all absolutely essential, features. Not everyone would agree, I know. I also never use the memories that all modern HF transceivers have, finding the band-stacking registers perfectly adequate, so to me memories are merely bells and whistles, but no doubt some operators find them essential.

## PRICE

A Rolls-Royce or Ferrari costs many times the price of a basic Mini or Hyundai, yet all will get you perfectly well from A to B. Some simply do it faster, more comfortably or in more style than others. It is much the same with HF transceivers: the top of the range transceivers cost several times the amount of the most basic. It could well be that price is the most important factor to you. If so, at the time of writing you can buy a new HF transceiver (the Alinco DX-SR9) for £499. If money is no object, you could spend £8900 on an Icom IC-7851. Most of us will probably be somewhere in between: we will want several features that the more basic transceivers do not include, and so we will be prepared to pay for them, but we may feel that we cannot justify the expense of buying the absolute top-of-the-range models.

*The Kenwood TS-930S, a great transceiver from the 1980s, can be picked up at bargain prices if you're lucky. If you can't justify the cost of a new top-end transceiver, a rig such as this might well be the solution.*

## NEW OR SECOND-HAND?

The question of whether to buy a new transceiver or a second-hand one is a thorny issue. If money is no object, then – certainly – buy a new rig. If you're really hard up, a second-hand transceiver may be the only realistic option. But what if you are undecided about whether to buy a modest new transceiver or spend roughly the same amount of money on a higher-specification second-hand model? That is where we hope this and the next chapter will prove to be helpful. Although older second-hand equipment may not have the wide variety of features a present-day transceiver possesses, its basic performance may be *better* than some modern-day equipment.

## NOISE BLANKER

This extremely important feature is, surprisingly, often completely overlooked when considering the purchase of a new transceiver. (Note that here we are discussing the Noise Blanker, abbreviated NB and not Digital Noise Reduction, DNR, which is covered later.) If you happen to live in an area with a high local noise level the best receiver in the world can be rendered almost useless if its noise blanker is ineffective. If you *do* live in a noisy area – and sadly this is becoming more and more common wherever you live in the world – it is essential that your transceiver's noise blanker can eliminate, or at least reduce, this noise level.

Unfortunately, while most are good at eliminating the noise generated by car ignition systems, not all noise blankers work well on other types of noise, such as that often radiated by overhead power lines, LED lights, plasma TVs etc. If at all possible, therefore, try to check out a couple of different transceivers' noise blankers at your own location before buying.

It should be noted that when the noise blanker is switched in, the receive performance can deteriorate, often quite markedly. The effect of this is that strong signals close to the frequency being received – even if they are outside the receiver's passband – will appear to be 'wider' and sound as if they are splattering badly. This can cause a level of interference to the wanted signal that is simply not there with the noise blanker out of circuit. Many an operator has been accused of overdriving his amplifier and causing splatter when the real reason is that the use of a noise blanker has compromised the receiving station's performance! At the same time there will be noticeable distortion on the audio of the signal being received, particularly if that is itself a strong signal. All this may be a small price to pay, though, if using the noise blanker reduces noise from, for example, overhead power cables from S9 to a more manageable S4 or S5.

## DSP & DIGITAL NOISE REDUCTION (DNR)

DSP stands for digital signal processing and was originally used at audio frequencies (AF DSP) to provide noise reduction facilities, a notch filter and bandwidth filtering on receive only. Before DSP was offered in transceivers, outboard AF DSP units were available from the late 1980s onwards. Placed at the end of the audio chain, between the transceiver and loudspeaker (or, more likely, the headphones) these could be used to filter the received audio. The intermediate frequency (IF) filters in the transceiver were still conventional crystal or mechanical filters.

More recently, DSP has been employed at the intermediate frequency of transceivers (IF DSP) and these days it is used in most current transceivers to carry out the IF filtering, both on transmit and receive, and several of the receiver's functions including noise reduction and notch filtering. DSP notch filters can be useful by removing a single frequency heterodyne, or tone, that is causing interference to the wanted SSB signal. However, a manually-adjusted IF notch filter is generally at least as effective. In some transceivers, DSP is also used to put a variable-frequency peak or dip in the received audio (the so-called 'contour' control).

I may be in the minority here, but I feel that digital noise reduction (DNR) has something of the 'king's new clothes' syndrome about it. Everyone says how wonderful it is, but is it *really* as good as people make out? Sure, DSP noise reduction can reduce some random background noises, but it also reduces the level of the wanted signal by a similar amount and at the same time introduces distortion to the voice, sometimes severe distortion if the DNR control is advanced too far. The effect is often described as making the voice sound as if it is speaking from the far end of a long drain pipe. After 25 years of using equipment with DSP, I have not yet found a single instance of where DSP noise reduction, whether implemented at AF or IF, can make an unreadable signal readable.

Where IF DSP really comes into its own, though, is in the flexibility it provides for the transceiver's IF filtering, on both transmit and receive. On transmit, DSP can be used to tailor the transmitted audio by emphasising different parts of the audio spectrum, and to vary the overall width of the transmitted signal (both of these features are discussed in more detail in Chapter 5). On receive, a wide variety of bandwidths are made available, allowing the operator great flexibility when combatting interference.

## TWIN RECEIVERS AND 'DUAL WATCH'

Most transceivers have one receiver, as might be expected. But a growing number have two, and these transceivers are of particular interest to DXers. (Here we are not talking about two VFOs – all modern HF transceivers have two VFOs these days – but two quite separate, independent, receivers.)

Having twin receivers allows the operator to monitor two different frequencies simultaneously, with the audio from one receiver going to the left side of a pair of stereo headphones and the audio from the other going to the right side of the headphones.

What is the purpose of the second receiver? How is it used? Many operators may never find they need to use the second receiver and will quite happily make all their contacts with just the standard first receiver that all transceivers have. But for the DXer, dual receivers give them a distinct advantage. When a DX station has many callers, they often use a technique called 'split' operation, in which they transmit and receive on slightly different frequencies. Without twin receivers, you can listen *either* to the DX station *or* the stations calling him (the pile-up), but you cannot listen to both at the same time. With twin receivers you can, and this gives the operator with twin receivers quite an advantage over those who do not have this facility when chasing DX stations. More about split frequency operation in Chapter 7.

An alternative facility to dual receivers, but very much a second best, is known as 'dual watch'. Here the two receivers are not completely independent of each other and the audio of the two receivers is combined, allowing two frequencies to be heard together. Separate AF Gain (volume) controls or a 'balance' control adjusts the relative volume of audio from each receiver, but the lack of the stereo spatial separation of true twin receivers makes this technique less effective.

Ask DXers who have upgraded from a transceiver with a single receiver to one with two receivers and invariably they will say that they would not go back to a single receiver. As a DXer, it is therefore definitely worth considering buying a transceiver with two receivers.

Further details of the features provided by various transceivers can be found in *The Rig Guide* **[3]**, published by the RSGB.

The remaining transceiver features to be discussed are concerned with the transmission of SSB speech and that warrants a separate chapter.

## REFERENCES

[1] *Hart Reviews – The Best of RadCom Equipment Reviews*, Peter Hart, G3SJX, RSGB 2015, available from www.rsgbshop.org
[2] GQRP Club: www.gqrp.com
[3] *The Rig Guide*, edited by Steve White, G3ZVW, RSGB (updated regularly).

# 5 Transmitting SSB Speech

NOT SO MANY YEARS AGO, you would buy your transceiver, plug in a microphone and antenna, switch on and start making contacts. You can still do that of course, but many modern transceivers offer a bewildering variety of features that need to be set up in order to get the best out of the set. Most of these are intended to tailor the transceiver to your particular operating habits, voice characteristics etc, and are 'set and forget' controls, i.e. once adjusted you will only rarely need to touch them again.

## SSB SPEECH QUALITY

If you tune across the SSB part of any band you will hear signals which vary widely in quality, from the outstanding to the frankly dreadful. Why is there such a wide range of SSB sound quality? There could be many reasons: the transceiver itself, the microphone, whether or not the transmitter is being overdriven, speech processing, the width of the transmit filter, the transceiver's internal audio tailoring, the use of an outboard audio equalisation unit and so on. In this section we will look at the ways in which you can have a superb-sounding SSB signal – and the ways in which you can destroy perfectly good audio and make it sound terrible (or rather avoid doing so).

In the same way that hi-fi audiophiles reckon that valve audio amplifiers always produce a better, more rounded, 'fuller sounding' quality than newer transistorised amplifiers, so there are those amateurs who believe that old valve transmitters produce better-sounding SSB than their newer solid-state counterparts. There could be something in this, although the reason may have nothing to do with the use of valves but instead perhaps be because older transmitters tended to use wider SSB filters. Anyway, these days – unless you are going to home-build your own transmitter or transceiver using valves – there is no choice in the matter since all current HF SSB transceivers are solid state.

That is not to say that modern transistorised SSB transceivers all sound alike. The basic transmitted audio quality of transceivers does vary from manufacturer to manufacturer and even from model to model. Indeed, the ability to tailor the transmitted audio of many modern transceivers means that it is possible to make their transmission quality very poor if you do not know what you are doing – more on this later in the chapter. Nevertheless, most (though not all) transceivers allow for very good SSB speech quality if they are set up or adjusted sensibly.

Assuming the transceiver itself is capable of providing good quality transmitted audio on SSB, the first stage to look at is the microphone.

## MICROPHONES

These days, microphones are generally supplied with new transceivers. Most manufacturers use dynamic microphones; the exception being Icom, which uses electret microphones for their transceivers.

The microphones that come with most transceivers can provide perfectly acceptable (although not usually outstanding) audio quality. In almost all cases, they are hand (or 'fist') microphones, with a press-to-talk (PTT) switch on the side. These are fine for casual operating, but their use becomes awkward if you are trying to adjust the controls on the transceiver at the same time as speaking, and especially if you use computer logging and are attempting to use a keyboard while talking.

An alternative is the desk microphone, but if anything their use is more awkward still, as it is rarely possible to have them at the correct height for speaking into if they are sitting on a desk as designed. As a result, most operators end up holding their desk mics, and these are always larger and heavier than the equivalent hand mic.

The stock mics supplied with the transceiver are not necessarily made by the equipment manufacturers themselves; in most cases they are OEM devices and simply 'badged' with the familiar names. Even when manufacturers use similar mics, the plugs are wired differently, so an otherwise almost-identical Yaesu mic cannot be used on a Kenwood transceiver, or vice versa.

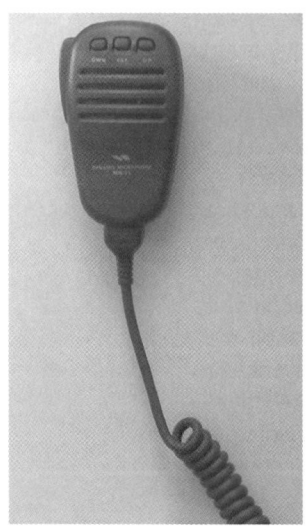

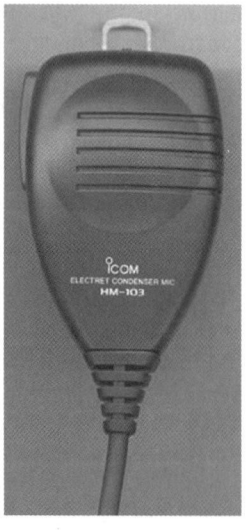

In some cases these microphones, or the inserts in the mics, are not designed primarily for radio transmission use but instead for karaoke (a vast market, especially in the Far East), public address or paging announcements. Even when they are designed for radio, they might be intended for FM communications rather than SSB, where it is desirable to have a peak around 2kHz. Also, the mass production techniques used to produce hundreds of thousands of microphones at a cheap price simply do not allow for the precision required for high-quality audio.

*Left: The Yaesu MH-31 dynamic microphone, as supplied with many Yaesu transceivers (such as the FT-857D). Right: The Icom HM-103, an electret condenser microphone, used with (for example) the IC-706 series of transceivers.*

The result is that, on SSB, many stock mics tend to sound too bassy for

## 5 — TRANSMITTING SSB SPEECH

the average Caucasian male voice. If your microphone is too bassy, one simple way of cutting its bass response is to put a capacitor between the mic itself and the mic socket on the transceiver. A switch allows the capacitor to be bypassed when necessary – see **Fig 5.1**. Not all mics (or voices) require this: depending on your own voice characteristics, the microphone used and the transceiver, you may well find that a stock microphone works well for you. I used a Kenwood TS-930S for many years and received nothing but positive comments on my transmitted audio when using the Kenwood MC-43S hand microphone.

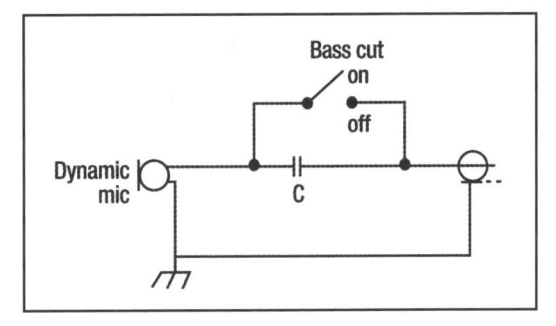

*Fig 5.1: Adding a capacitor can decrease the bass response and improve the communications quality of a microphone. The switch simply allows the bass-cut capacitor to be bypassed when required.*

### ENTER BOB HEIL

Bob Heil, K9EID, worked as an audio engineer with many of the big name rock bands of the 1970s. When it came to amateur radio SSB operation he disliked the audio quality provided by many microphones, which he described as "mushy". He developed his own microphone insert, which can be used as a 'drop-in' replacement for many of the stock dynamic microphones. The result was the Heil [1] HC-5, a ceramic microphone insert with an audio peak between 1.5 and 3kHz (+6dB at 2kHz), providing clear and 'pleasant' sounding audio quality with most voices. To cater for the DXer and contester, Heil also developed the HC-4 insert, which has a stronger peak than the HC-5 (+10dB at 2kHz), and at higher frequencies: between 2 and 4kHz. This provides 'toppy' sounding audio that really cuts through the QRM. As Heil himself says, "not pretty sounding – but in your face extreme audio". It may not sound very attractive for local communications, but it is what is required for weak-signal DX working and contest operating. **Fig 5.2** shows the frequency responses of the HC-5 and HC-4 microphone elements.

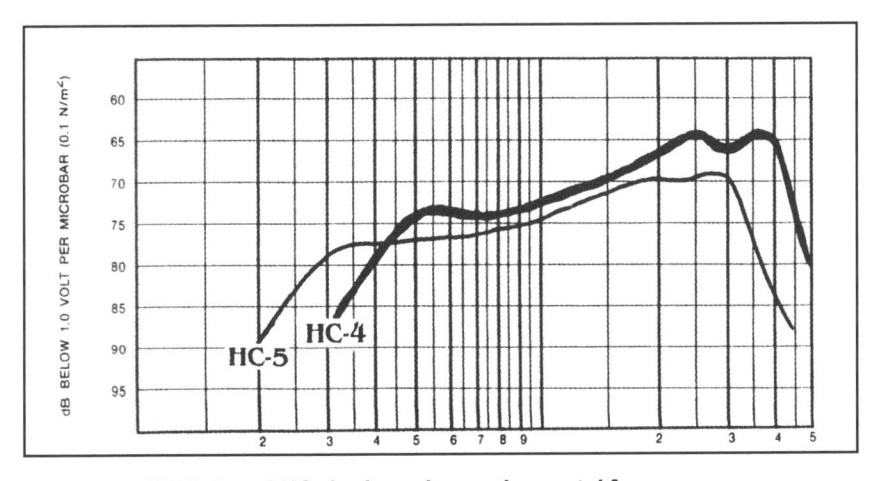

*Fig 5.2: Heil HC-5 and HC-4 microphone elements' frequency responses.*

# HF SSB DX BASICS

Heil also developed a series of headsets – headphones with boom microphones attached – using either the HC-5 or HC-4 mic insert. These proved very popular with SSB operators and particularly among DXers, DXpedition operators and contesters because their use, with a footswitch replacing the PTT switch on the microphone, allowed both hands to remain free to use a computer keyboard for logging and to operate the transceiver.

The Heil Pro-Set Plus headset has both the HC-5 and HC-4 microphone elements built in, and user-selectable by means of a small switch on the headset's boom. It is interesting to note that off-air reports are quite different from any conclusions that may be drawn by monitoring one's own voice using the 'Monitor' facility in many transceivers, where often little difference is evident when switching between the two microphones.

The Heil Sound website has studio recordings made using various Heil microphones for comparison purposes [2].

## OTHER MICS

Heil might now be the best-known company producing replacement microphones for amateur SSB transceivers, but it is not the only one. Founded as far back as 1925, Shure [3] has been producing microphones suitable for amateur radio use for decades. Current models include the Shure 522 and 527B. The 527B has a sharp peak around 3kHz (**Fig 5.3**) making it particularly suitable for DX work on SSB.

Another manufacturer is Adonis [4]. Some of their mics, such as the AM-708E, have a built-in speech compressor. The three most popular Adonis microphones for amateur radio use are the AM-308, AM-508E and AM-708E. Note that these three microphones all have electret inserts with built-in pre-amplifiers. As such, they need to be powered when in use, either with batteries or from an external DC power supply.

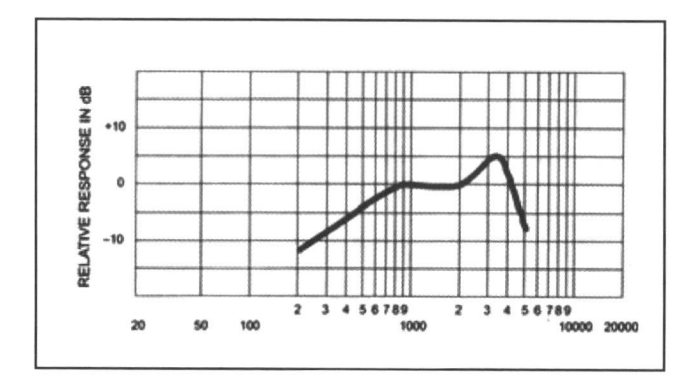

*Fig 5.3: Frequency response of Shure 527B mic.*

## ALC

No matter how good the microphone – or the transceiver for that matter – an SSB transmission can be rendered almost unintelligible simply by the transmitter being overdriven.

A good way to prevent this from happening is by always keeping one eye on the ALC (automatic level control) meter of the transceiver. If you exceed the range shown on the ALC meter, your signal does not get any stronger, nor your audio any 'louder'. Instead your signal becomes distorted and wider, and so causes unnecessary interference to other stations operating close to your frequency.

# 5 — TRANSMITTING SSB SPEECH

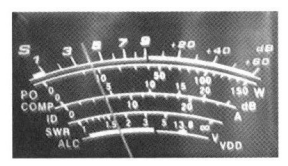

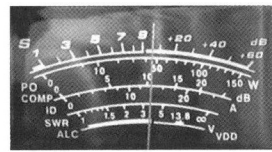

*The ALC meter is the bottom scale of this multi-scale meter. Left: no ALC. Centre: middle of the ALC scale: this is approximately the level one should be aiming for on speech peaks. Right: maximum ALC: danger of overdriving. Above this level the transmitter will certainly be overdriven, causing distorted audio, splatter and therefore interference to other band users.*

A few transceivers (e.g. the Icom IC-7200) have a VOGAD (Voice Operated Gain Adjusting Device) circuit included in the transmit audio chain. A VOGAD is a kind of automatic gain control (AGC) circuit in the microphone amplification stage. It is used to reduce the signal's *dynamic range*, thus increasing the average transmitted power and preventing overmodulation. This may sound similar to the action of a speech compressor but the advantage is that the VOGAD optimises the level of modulation in real time, before the conventional speech processing stage. A VOGAD circuit has a fast attack but a long delay time, so that any background noise, breathing sounds etc are not heard between words.

## SPEECH PROCESSOR

Those operators who only ever take part in local 'natter nets', where signals are S9 or more, may not want or need to use speech processing, but for all SSB DXers some form of speech processing or clipping is virtually essential. A correctly set up speech processor can increase the readability of the signal and increase your signal strength at the far end.

Almost all modern rigs have a speech processor built in. It is a pity that speech processing has become something of a rude word among many operators, who therefore refuse to use this important and useful facility. The reason for this is that so many operators *abuse*, rather than use, speech processing, and overdrive their transceivers, causing distorted signals and splatter, which in turn causes unnecessary interference to other band users.

The trick is to learn how to use the speech processor properly — how to set the levels correctly, and then leave them alone!

In its simplest form, speech processing works by both compression and limiting, decreasing the dynamic range of the voice: the quieter sounds are made louder (decreasing or *compressing* the audio dynamic range) and the louder sounds are *limited*, so that they do not exceed a certain level.

Without a speech processor in circuit, the transceiver's microphone gain control would have to be set so that the *loudest* of speech peaks did not overdrive the transceiver, thus causing distortion and splatter. In an ideal world, the loudest speech peaks would equate to the transceiver putting out its maximum power, say 100W. However, most of the speech would be at a *much* lower level than this and therefore most of the time the transceiver

# HF SSB DX BASICS

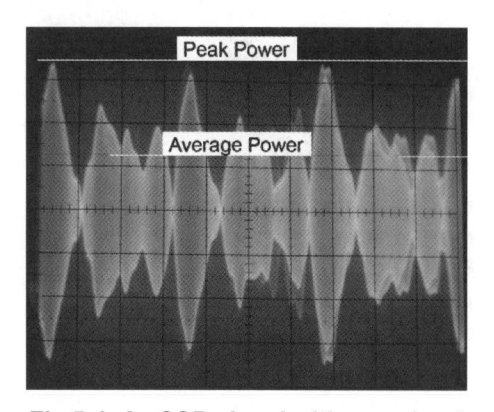

*Fig 5.4: An SSB signal with a moderate amount of speech processing.*

would be radiating a much lower amount of power than 100W. With the processor in circuit, the maximum speech peaks are limited so they do not overdrive the transceiver, but the quieter sounds are brought up much closer to this level. **Fig 5.4** shows an oscilloscope display of an SSB signal, with the average and peak power levels marked. When the speech processor is in circuit, the average power level of the transceiver is much closer to the maximum peak power output of the transceiver (i.e. 100W PEP in this example) than it is when the speech processor is not being used.

There are two main types of speech processing: AF and RF. In AF speech processing, as the name suggests, the processing takes place at audio (voice) frequencies. The problem with this technique is that the clipping creates harmonics and mixing products which cause distortion to the audio. It is impossible to remove all of these harmonics by filtering them out, because some fall within the normal SSB bandwidth of 300 – 2700Hz. For example, an important part of the human voice spectrum is around 500 – 1500Hz. There will be harmonics of the 500Hz part of the spectrum at 1000, 1500, 2000 and 2500Hz, while the second harmonic of the 750Hz part of the voice will fall at 1500 and 2250Hz and so on. The 2250Hz harmonic will mix with the one at 1000 to produce a mixing product at 1250Hz. Meanwhile, the second harmonic of the 1000Hz part of the voice spectrum will be at 2000Hz and this will mix with the other harmonics. The net result of all the harmonics mixing is that the signal may sound 'loud', but it will also sound distorted – horribly so if the level of clipping or compression is set too high.

In the case of RF speech processing, the audio is translated to much higher radio frequencies where the processing is done, and then translated back to audio. In this case, any harmonics generated by the clipping are at multiples of the RF frequency at which the clipping takes place and so can be removed with an SSB filter. A simplified block diagram of an RF speech clipper is shown in **Fig 5.5**.

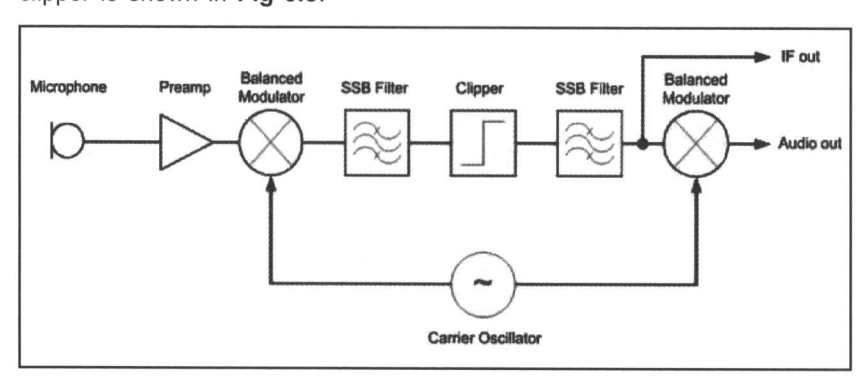

*Fig 5.5: Block diagram of a basic RF speech clipper.*

## 5 — TRANSMITTING SSB SPEECH

It is these unwanted harmonics that cause the distortion: remove them and the resulting signal sounds much 'louder' but without appreciable distortion. RF processing is thus superior to AF, although it has to be said that very satisfactory results can also be obtained by AF speech processors provided the level of clipping used is not too great. For an excellent and detailed description of how speech clipping works, see the article by Ian White, G(M)3SEK [5].

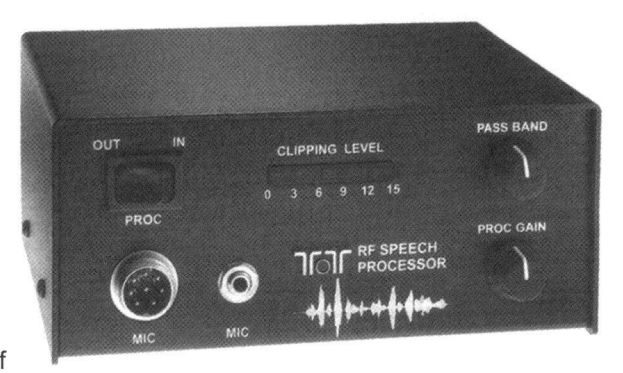

*The Ten-Tec 715 RF speech processor.*

Pretty much every transceiver available today has its own built-in speech processor, though some are more effective than others. If your transceiver's speech processor is not that effective, for example if it is an AF speech processor – or if you use 'classic' (old) equipment that does not have one at all – it is possible to add an out-board unit. The Ten-Tec 715 RF speech processor is one such unit. According to the manufacturers, it will boost the average output power by 6dB, equivalent to going from 100W to 400W out. That should give 1 or 2 S-points of gain at the receive end.

## TRANSMISSION AUDIO TAILORING

The advent of IF DSP has brought about the possibility of being able to tailor both the transmitted audio quality and the overall width and shape of the transmitted signal. Top-end transceivers, at least, now normally allow the operator to increase or decrease the relative level of the low, middle and upper range of frequencies of the audio chain.

Even some budget transceivers have a simple form of audio tailoring, be it a bass cut and / or top cut filter, which can be used to match more closely the operator's voice to the microphone used with the transceiver. The Yaesu FT-857D for example (a budget mobile transceiver and now a relatively old design), uses DSP to provide high cut (emphasising lower frequencies), low cut (emphasising higher frequencies) or high *and* low cut (emphasising mid-range frequencies). This provides a limited way of altering the frequency response of the transmitted audio by allowing you to roll off any excessively high or low frequency components in your voice.

Newer and more sophisticated mid to top-end transceivers, on the other hand, often provide a vast range of audio tailoring possibilities. Such equipment often has a 'parametric microphone equaliser' which allows the centre frequency of the low, middle and high range to be adjusted independently in (for example) 100Hz steps between 100Hz and 3200Hz. For each of the low, middle and high range of frequencies, the relative level can be adjusted up or down. The Q-factor (width) of each can also be adjusted. Together with the possibility of changing the actual bandwidth of the transmitted signal (see later), there are hundreds of thousands of possible permutations, all of which will make the transmitted audio sound a little different.

# HF SSB DX BASICS

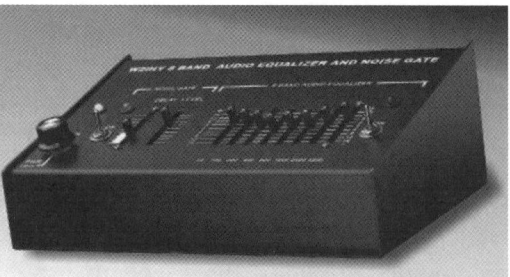

*Above: The '8 Band EQ' and the 'EQplus', both by W2IHY Technology Inc. Right: the MFJ-653 'Speech Articulator'. All three units can be used to improve the quality of your transmitted SSB audio.*

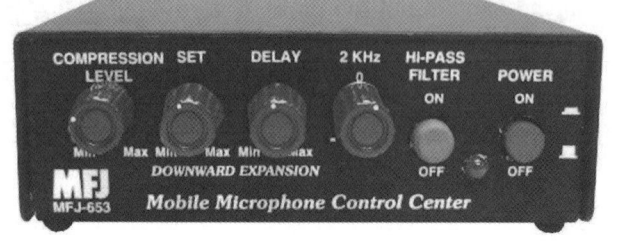

The bad news is that extreme settings will certainly make the audio sound extreme – extremely poor! Many operators adjust these settings only to come to the conclusion, after much time and effort, that the best-sounding audio is produced using the default settings – the ones the rig comes with out of the box!

If your transceiver does not have transmission audio tailoring built in, it is possible to buy outboard units to do the job. W2IHY Technology Inc **[6]**, run by Julius Jones from Staatsburg, New York, produces a range of audio equipment for the SSB operator. The two best-known units are the '8 Band EQ' and the 'EQplus'. The 8 Band EQ is an equaliser with eight slider controls to adjust the low, mid and high components in the audio. It includes a noise gate to reduce background noise. The EQplus has dual band equalisation, a 'Down-ward Expander' for noise reduction and it includes a compressor and limiter to increase 'talk power'. Either can be used as a stand-alone unit or the two can be used together for greater flexibility. Both have a universal interface to allow the use of almost any microphone with any transceiver.

A unit produced by MFJ **[7]**, the MFJ-653 'Mobile Microphone Control Center', also known as the 'Speech Articulator', has a different function. It boosts the microphone's audio frequencies around 2kHz by up to 16dB and cuts the low frequencies that provide little intelligence, thus improving transmit speech intelligibility and providing 'punchier' audio quality. Judging by its name, this unit was obviously originally designed for mobile operators but it looks like it would also be a useful piece of equipment for DXers who do not have audio tailoring built into their transceiver. The MFJ-653 also incorporates a speech compressor and a noise gate to remove background noise during the pauses in speech. Input microphone sockets include an 8-pin round, an RJ45 modular, and a programmable 3.5mm jack, allowing it to be used with almost any microphone, whether high or low impedance and dynamic or electret – even cheap computer headsets.

## TRANSMISSION BANDWIDTH

Many of the newer and more sophisticated transceivers allow the operator to select the actual bandwidth of the transmitted SSB signal. The bandwidth of an SSB signal is normally around 2.4kHz, with audio frequencies of typically 300 – 2700Hz being transmitted.

Taking the Yaesu FT-2000 as an example, in addition to the default of 2.4kHz (300 – 2700Hz), a narrower setting of 2.2kHz (400 – 2600Hz) can be selected, or no fewer than four wider settings. These are 2.6kHz (200 – 2800Hz), 2.8kHz (100 – 2900Hz) and 2.95kHz (50 – 3000Hz), plus a special so-called 'hi-fidelity' setting whereby the transmitted bandwidth is in excess of 3kHz. An audiophile's definition of 'hi-fi' would be accurately-reproduced audio over a bandwidth ranging from about 20Hz to 20kHz, so clearly no amateur SSB transmission can even approach *true* 'hi-fi' quality. However, in relative terms, an SSB transmission which is 3kHz wide or greater is capable of reproducing the human voice more accurately than one restricted to just 2.4kHz, so there is some justification in using the term.

What is the effect of changing the transmitted bandwidth – what does the SSB signal actually *sound* like after these changes have been made?

The narrower setting of 2.2kHz limits the lowest audio frequency being transmitted to 400Hz and the highest to 2600Hz. This restricted bandwidth has been described as sounding like a tinny telephone line and most people tend to dislike the sound of the audio thus produced. There is, arguably, a small advantage when the signal path is very marginal, though. Because all the available power (e.g. 100 watts) is concentrated into just 2.2kHz, instead of being spread out over 2.4kHz or more, the signal strength of the narrower signal will be slightly greater.

However, those operators who are able to adjust their transmitted bandwidth generally opt not for the narrower setting but for one of the wider ones. Wide-band SSB is sometimes referred to as 'Extended SSB' ('eSSB'), but when does 'standard SSB' ('sSSB' perhaps?) become eSSB? John Anning, NU9N **[8]** is a staunch advocate of eSSB transmission and by his definition it is any SSB transmission with a bandwidth of 3kHz or more. He explains, "The reason for this is that high-frequency audio from 3kHz and above starts to support a significant difference in clarity, 'openness' and fidelity of the audio signal that better reproduces natural energy found in the human voice. Even though vocal chord energy diminishes rapidly above 3kHz, the all-important high-frequency consonants of human speech such as the 'S', 'T', 'SH', 'CH', 'K' and 'Z' sounds that are formed with various combinations of the tongue, roof of the mouth and teeth are well above 3kHz. The accurate reproduction of these sounds is essential for high-definition speech with less listener fatigue." By NU9N's definition, the FT-2000's wider bandwidth settings of 2.6, 2.8 and 2.95kHz are not eSSB but simply wider than normal – though still standard – SSB transmissions.

The change from the default 2.4kHz to a slightly wider 2.6kHz bandwidth provides a subtle change in the audio quality which many operators might prefer. It's true that it does provide 'smoother', more 'rounded', sounding audio, but it is less effective at cutting through interference than the 2.4kHz setting, so is less use for DXers or contest operators. However, this might be considered a small price to pay for those operators who normally enjoy chatting with local stations at good signal strength levels, and only occasionally chase DX stations or go in for contest operation. Besides, it is not a difficult operation to return the transceiver to its default setting when required.

It should be noted, though, that if the station receiving a wide-band 'hi-fi' SSB transmission is receiving it in a standard 2.4kHz bandwidth, any advantage will be lost; the receiver's bandwidth needs to be set to match the width of the transmitted signal. In practice this is easy to do with transceivers that use IF DSP for their filtering as there are likely to be a wide range of IF filter bandwidth settings available to the operator.

One problem, though, is that under crowded or noisy band conditions, opening up a receiver's bandwidth to 3kHz or greater is likely to let in adjacent-frequency interference as well as more local noise such as the pops and crackles that are, sadly, normal in most urban and suburban locations. So a really wide-band 'hi-fi' SSB signal can only be received under quiet band conditions and, preferably, with strong local signals (which begs the question "why not use FM on VHF / UHF instead?")

Those operators who wish to transmit a truly high-quality SSB signal usually do not stop at adjusting the transceiver, but also use wide-range broadcast-quality microphones and outboard audio processing equipment too, Behringer [9] being a particular favourite brand.

The results can, no doubt, be astounding, but they can also be abysmal if the transceiver and peripheral equipment are not set up well. Unfortunately, some amateur radio operators appear to aspire to be 'broadcast announcers' with booming bass voices. Since they do not have a natural booming bass voice, they set their transceivers to the widest possible setting (as this increases the low frequencies transmitted as well as the high frequencies), then adjust the audio tailoring to emphasise the lowest frequencies possible. The result can be virtually unintelligible: on one evening I was called by a station whose signal strength was S9+20dB yet I could hardly understand a word being said. There was no 'top' or even middle element to his transmission, just a bass rumble each time he spoke. This is clearly not what eSSB should be all about, yet it can be heard on the air.

## DIGITAL VOICE KEYER

The digital voice keyer (DVK), also known as a digital voice recorder, is generally thought of as a requirement for serious contest operators – but not so for DXers. Nevertheless, they can be a very useful feature for SSB DXers, as we shall see. (Note that the DVK has nothing at all to do with the DV digital voice mode that was discussed in Chapter 1.)

DVKs allow the operator to record several messages that are repeated frequently, such as a CQ call, thus saving the voice. For many operators the DVK will fall into the 'Bells and Whistles' category, but a top contest operator can make several thousand QSOs in a weekend and so will have to call CQ, and give his callsign and the contest exchange many thousands of times. Having this automated is, in these circumstances, a real boon.

# 5 — TRANSMITTING SSB SPEECH

DVKs are found in many top-end transceivers. A few add-on units are also available, such as the MFJ-434B [7] pictured here.

So how is a DVK used by the SSB DXer? When attempting to make contact with a rare DXpedition station it is not uncommon to have to call for a long period of time – many minutes or hours (or sometimes off and on even for days!) – before the desired contact is made. Instead of having to give your callsign over and over again it is easier just to push a button on the front of your transceiver and have the DVK call the DXpedition instead.

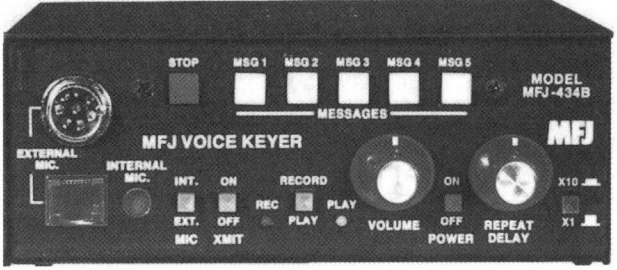

*The MFJ-434B Digital Voice Keyer.*

*Beware of overusing this facility, though:* 'continuous calling' is the bane of many DXpedition operators' lives and if you do this it *will* make you very unpopular. Since it is easier just to push a button than deliberately to transmit your callsign by using your voice, many would-be DXers get 'carried away' and call using the DVK when they should not be transmitting at all. (Naturally you should *only* call when the DXpedition station is specifically listening for callers and *never* when he is in contact with another station, or trying to receive the callsign of a station that is clearly not you!)

Most DVKs will also record the transceiver's *received* audio, usually on a continuous 'loop' system, for a period of, say, 30 seconds. When the operator wishes to hear something again he can stop the recording and play it back as many times as they wish. This facility is also useful for DXers, especially when conditions are marginal, as it allows the DXer to confirm to themselves that the DX station did indeed respond to them and that it was a 'good QSO'. Some DXers keep their best contacts or the rarest DX QSOs in the form of digital audio files stored on their computer – these can be fascinating to listen to years later and provide a much more vivid historical record than just a logbook entry or even a QSL card.

Finally, DX stations sometimes use DVKs to 'broadcast' frequently-sent pieces of information, such as QSL information, listening frequencies, or perhaps name and QTH. One Far Eastern operator I know sometimes conducts virtually the whole of his SSB QSOs using the various memory buttons of his digital voice recorder, with a CQ call on memory 1, name and QTH spelled phonetically on memory 2, QSL information on memory 3, and "QRZ?" plus his callsign on memory 4. The only time he actually uses his voice is to give the callsign of the station he is working. It may be argued that this is taking station automation a step too far, but of course this use of a DVK is perfect for those operators who have speech difficulties (or perhaps have simply 'lost their voice') but who still wish to operate on SSB.

## MONITOR FACILITY

Like the noise blanker discussed in Chapter 4, the Monitor facility is an oft-overlooked facility that can be very important for the SSB DXer. It allows you to hear your own voice as you are transmitting so as to get an indication of your outgoing transmitted audio quality. Suffice to say that (in my opinion, anyway) the Monitor facility is, if not an absolutely essential feature, then at least a very important one for the SSB operator.

Because the Monitor facility allows you to hear your own voice when you are transmitting this necessarily means that you must be wearing headphones, so as to avoid audio feedback or 'howl-round' caused by the microphone picking up sound from the loudspeaker and retransmitting it. It is anyway highly recommended always to wear headphones for all but the most casual 'armchair copy' contacts.

Being able to hear your own voice has three advantages. Firstly, you will immediately be able to tell if something is amiss with your transmitted audio, such as a loose microphone connection causing intermittent loss of transmission, or maybe RF feedback caused perhaps by a poorly-soldered PL-259 antenna plug or a poor or non-existent earth connection.

Secondly, if you are wearing headphones but have no 'sidetone' provided by the monitor facility you are inclined to shout. This is bad news: it makes your transmitted audio difficult to understand at the far end, you are more inclined to overdrive the transceiver if you are shouting, causing further distortion and probably splatter, and if you are transmitting for long periods of time, e.g. during a 24-hour or even 48-hour contest, your voice will eventually give out and you will end up with a sore throat or possibly even lose your voice altogether.

Finally, to a limited extent anyway, you will be able to make adjustments to your own transmitted audio, such as adjusting the speech processor settings or your transmitted audio bandwidth, and judge the results yourself in your headphones. I say "to a limited extent" because the monitor facility rarely gives a true reflection of the real transmitted audio quality and, besides, it is very difficult to make a fair assessment of your own voice when you can also hear yourself speaking, through bone conduction in your head. You really need someone else to give critical comments. Nevertheless, the monitor facility is a good start and allows you, for example, to rule out extreme settings that you can tell sound unacceptable.

## REFERENCES

[1] Heil Sound: www.heilsound.com
[2] Heil microphone comparisons (studio recordings): www.heilsound.com/sound_comparisons.htm
[3] Shure microphones: www.shure.com
[4] Adonis microphones: www.adonis.ne.jp/e-home.html
[5] 'Loud and Clear' ('In Practice'), Ian White, G3SEK, *RadCom*, May 2003.
[6] W2IHY Technologies: www.w2ihy.com
[7] MFJ Enterprises, Inc: www.mfjenterprises.com
[8] Extended SSB (eSSB), John Anning, NU9N: www.nu9n.com
[9] Behringer audio processing equipment: www.behringer.com

# 6 Propagation & SSB DX on HF

THERE ARE MANY BOOKS written about propagation and a short book such as this cannot even scratch the surface of what is a very wide-ranging subject in itself. So rather than go into any great detail, I will try to give an indication of what DX can be worked on each band and when, touching on how the solar cycle affects each of the bands and briefly mentioning the various modes of propagation involved.

All of the HF bands are affected in one way or another by the 11-year solar cycles (**Fig 6.1**). As this is being written, in late 2015, we are on the downward slope of solar cycle 24, which peaked in 2014, and therefore solar activity is declining, and will continue to decline for several more years. Cycle 25 is expected to start in about 2020 and peak around 2023 to 2025, but many experts expect it to be a very weak cycle, even smaller than the present cycle, which was itself the smallest since the invention of radio, as can be seen in the diagram below.

Most long-distance propagation on the HF bands is by refraction from the F2-layer of the ionosphere (see **Fig 6.2**). It is F2 propagation that is most affected by the solar cycles. Much simplified, the higher the frequency band, the more it is affected by the stage of the cycle and, generally, the greater the solar activity the better the propagation, particularly on the higher-frequency bands (14MHz and above).

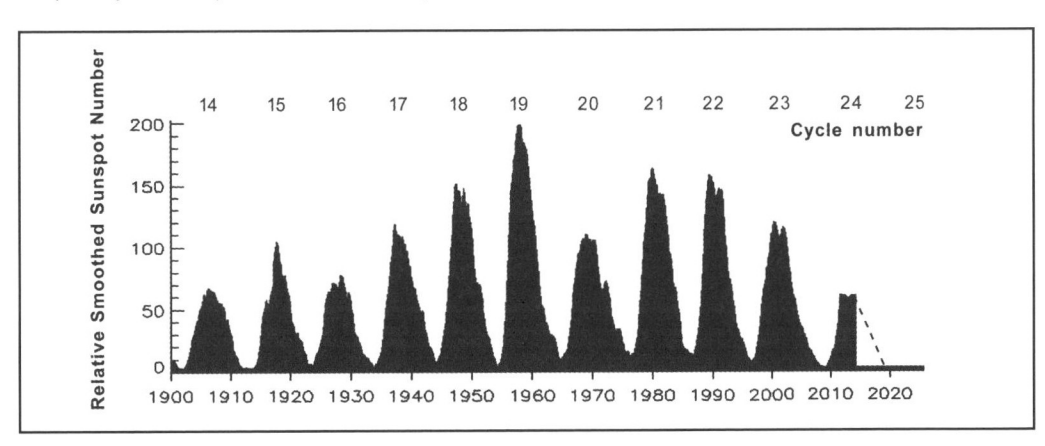

*Fig 6.1: Solar cycles since 1900.*

# HF SSB DX BASICS

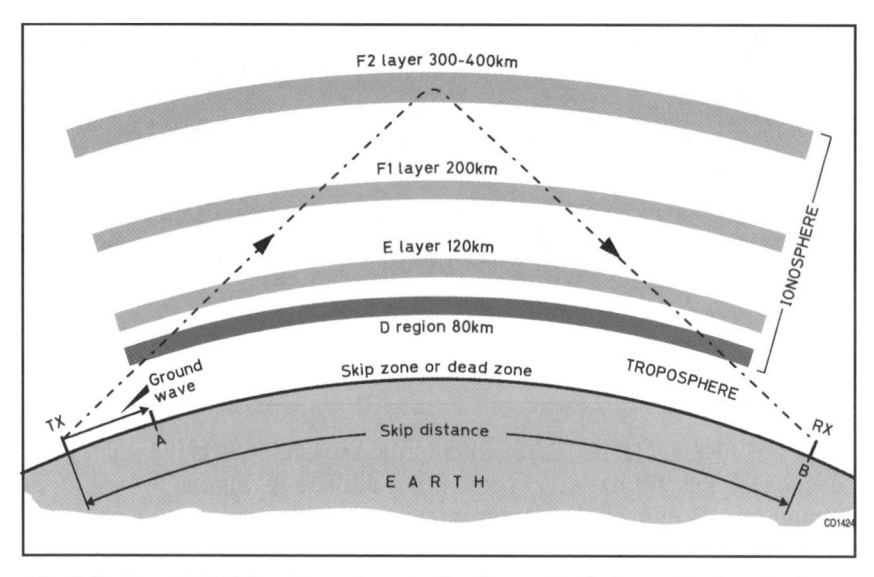

*Fig 6.2: Most HF DX – though not all – is worked through signals being refracted from the F2 layer of the ionosphere.*

F2 propagation tends to peak in the autumn (in the northern hemisphere mid-September to mid-November) and spring (March to mid-April) periods, with another peak in mid-winter (December), but be at its poorest during the summer months (May to September).

What all this means in practice is that, for the next few years (until Cycle 25 starts to peak), DX propagation is likely to be only fair to poor but with occasional 'bursts' of activity during the peak seasons mentioned above. However, while propagation during those 'bursts' may be relatively good, it is unlikely to be as good as during the glory days of 1999 – 2002 or even the most recent solar cycle peak of 2013 – 2014.

The other layers of the ionosphere shown in the diagram above have different effects on HF propagation. The D-layer is only present during daylight hours and dissipates shortly after dark, re-forming again soon after dawn. This layer is responsible for the *absorption* of radio waves, particularly below 14MHz, which is why DX contacts are generally only possible on 80 and 40m after dark. The E-layer can refract radio waves on 80 and 40m, but is mainly of interest during Sporadic E events, of which more later.

Now we'll take a look at each SSB band in turn, from 10 to 80 metres.

## 28MHz – 10-METRE BAND

The 10m band is the widest of the HF bands with no less that 1.7MHz of spectrum available: 28000 – 29700kHz. According to the latest RSGB Band Plan [1] which is based on the IARU Region 1 band plans, the 'All Mode' part of the 10m band (which obviously includes SSB) is from 28225 – 29700kHz. However, the SSB part of the band is *normally* considered to be 28300 – 29000kHz, with beacons operating below 28300kHz and AM and FM stations above 29000kHz.

# 6 – PROPAGATION & SSB DX ON HF

During the major contests at times of high solar activity the band may be so crowded that SSB stations will operate as low as 28225kHz and as high as 29200kHz or so.

However, at times of *low* solar activity far fewer stations will be heard and SSB operation then tends to be between 28400 and 28550kHz. When there is very little long-distance propagation on 10m almost all SSB activity is within a few tens of kilohertz of 28500kHz. DXpedition stations tend to use 28495kHz and generally listen 5kHz (or 5 to 10kHz) higher, whereas IOTA DXpeditions often use 28460kHz.

10m is the HF band most affected by solar activity, with excellent DX propagation available during the peaks of the solar cycle but long lean years during solar minima. If the prediction that Cycle 25 will be a very weak one comes to fruition, we may well be in for some comparatively slim DX pickings on 10m for the next six or seven years. I say "comparatively" because when 10m *is* open, strong signals can be heard from literally all over the world.

Even when the 10m band is otherwise 'dead', north-south propagation can occur. From the UK, stations in southern and central Africa, or Brazil, Uruguay and Argentina can sometimes be worked, often with good signals, when stations to the north of them are quite inaudible.

Although by definition certainly an HF band, in some ways propagation on 10m more closely resembles that of a low VHF band than that of the other HF bands. For example, *auroral* and *Sporadic E* propagation, almost always considered to be VHF phenomena, are in fact also present on 10m. Auroras tend to happen during periods of high solar activity and particularly for the two or three years *after* the peak of solar activity. A *radio* aurora can occur even when there is no visual aurora apparent, but if the 'northern lights' *are* being seen – typically in Scandinavia or Scotland – then there is a strong possibility of auroral conditions on the radio.

*A spectacular visual aurora on the Faroe Islands, 5 February 2011.*

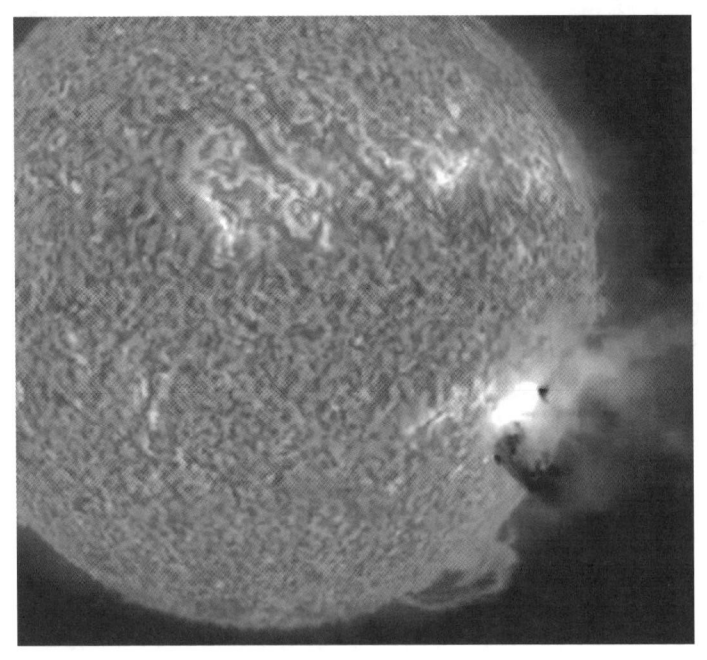

*A massive solar flare – the precursor to an aurora.*

Auroras occur after a solar disturbance such as a solar flare. Precisely how the aurora is formed is beyond the scope of this book, but is well covered in *Radio Auroras* by the late Charlie Newton, G2FKZ **[2]**.

On most of the HF bands the effect of an aurora is that signals are greatly attenuated and SSB signals have a 'fluttery' sound. This is particularly apparent on Scandinavian stations and especially those north of the Arctic circle. Particularly strong auroras can lead to an almost complete radio black-out, especially at higher latitudes.

However, in common with some of the VHF bands, on 10m contacts can be made via the auroral curtain with stations in northern parts of Europe such as Scandinavia, Poland, the Baltic States and north-western Russia.

10m also exhibits *Sporadic E* propagation (abbreviated Es) during the summer months (typically May to July) throughout the solar cycle. By its very nature, this type of propagation *is* sporadic and may occur on one day but not the next. From the UK, Sporadic E propagation will bring in strong signals from around Europe and perhaps North Africa or the closer parts of the Middle East (Turkey, Cyprus, Israel, Jordan etc). Occasionally double hop or multiple hop Sporadic E might allow for openings to the eastern part of North America or the Caribbean.

## 24.9MHz – 12-METRE BAND

The 12m band, 24890 – 24990kHz, is a narrow band which sees considerably less activity than the adjacent 10m and 15m bands. SSB stations operate between 24930 and 24990kHz and DX can be found anywhere in that 60kHz.

Along with 17m, by international agreement there is no contest activity in the 12m band.

Propagation is very similar to that on 10m except that, during periods of marginal openings, 12m is more likely to be open than 10m. It is therefore worth checking the band and perhaps trying a CQ call even if both 10m and 12m appear to be 'dead'.

Like 10m, F2 propagation on 12m is mainly during daylight hours. That's not to say that contacts cannot be made after dark – they certainly can – but then they will usually be made in the direction of the sun. For example, after dark in the evening contacts will be made towards the west, where it is still daylight. During periods of high solar activity, 12m may stay open until well after dark, typically with signals from the Americas. However, during periods of low solar activity 12m may well close quickly after sunset and at the solar minimum it might not open at all for days at a time.

As with 10m, Sporadic E contacts around Europe and further afield can often be made on 12m during the summer months.

## 21MHz – 15-METRE BAND

The 15m band is much wider than 12m, at 450kHz (21000 – 21450kHz). SSB stations operate between 21150 and 21450kHz although there is also occasional AM and FM activity above 21400kHz. No US stations are allowed to operate on SSB below 21200kHz which means that 21150 – 21200kHz is often used by stations in the western hemisphere (Canada, Central America, Caribbean etc) and sometimes the Pacific, in order to avoid interference to or from USA stations. DXpeditions typically use 21295kHz and IOTA DXpeditions 21260kHz.

Being a wide band, 15m is rarely overcrowded, except during the major contests at times of solar maximum. During solar maximum, and for a year or two before and after, 15m is an excellent DX band. However, at solar minimum and for a couple of years either side of the minimum, 15m can often appear to be dead. When there is little activity on 15m, most stations tend to operate between about 21210 and 21310kHz.

Like 10m and 12m, propagation on 15m is mainly during daylight hours but it tends to stay open later in the evenings than both 10m and 12m and, during periods of high solar activity, it may remain open until midnight or 1.00am local time, with strong signals from Central and South America, the West Coast of North America, and – when conditions are particularly good – Hawaii and elsewhere in the Pacific. However, in contrast at solar minimum 15m might appear to be 'dead' for several days at a time.

## 18MHz – 17-METRE BAND

This is another narrow band, only 100kHz wide, between 18068 and 18168kHz. SSB activity is to be found between about 18115 and 18168kHz and DX stations operate anywhere in that part of the band. There is no contest activity in the 17m band, making it a 'refuge' for those who dislike contests.

17m is a popular band, and justifiably so because it is far less affected by low solar activity than are 10, 12 and 15m. Therefore there is usually some DX to be found on the band, even when the higher frequency bands are apparently dead. However, because of its popularity and because the band is so narrow it can often get crowded, making it difficult to find a clear frequency.

Like 15m, 17m can stay open till late at night during periods of high solar activity, but it is likely to close not long after sunset during low solar activity years.

# HF SSB DX BASICS

## 14MHz – 20-METRE BAND

20m is *the* quintessential DX band. 350kHz wide, from 14000 to 14350kHz, SSB activity can be found from 14112 to 14350kHz. The US General licence class (roughly equivalent to the UK Intermediate class) is allowed to operate from 14225 to 14350kHz with a power limit of 1.5kW, so this is a good part of the band to work US stations. However, within the USA the General part of the band is mainly used for domestic contacts and numerous US nets, so most (though certainly not all) DX activity takes place below 14225kHz.

14190 – 14200kHz is listed as "Priority for DXpeditions" in the RSGB Band Plan (**Fig 6.3**) and the IARU Band Plan, although in practice DXpedition stations also often use other frequencies: 14145 and 14185kHz are particular favourites. IOTA DXpeditions tend to operate on, or close to, 14260kHz. Because no USA stations are allowed to operate on SSB lower than 14150kHz, 14112 – 14150kHz is often used by stations in Canada and the Caribbean when they wish to work DX outside the North America area.

14225 – 14235kHz in the SSB part of the band is shared with SSTV (slow scan television) operators, so it is best to avoid those frequencies when operating on SSB.

DX can be found on 20m throughout the solar cycle and at all times of year, although the spring and autumn seasons provide definite peaks in propagation. Conversely, during the summer months, and particularly during

## RSGB Band Plan 2015

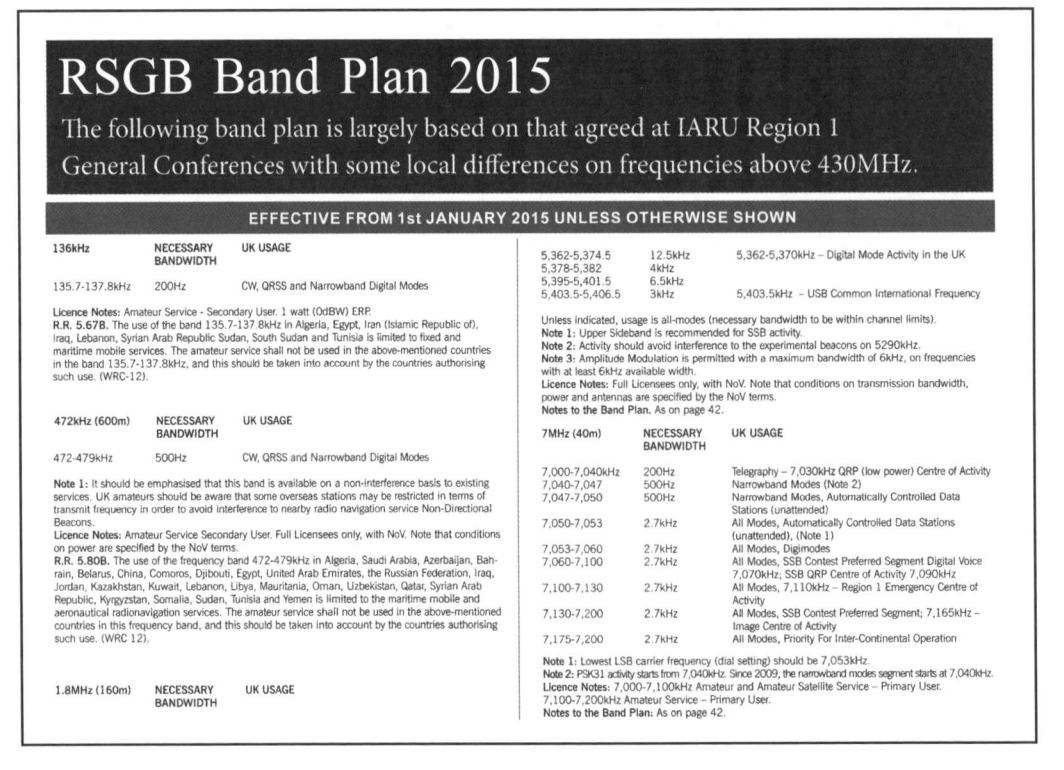

The following band plan is largely based on that agreed at IARU Region 1 General Conferences with some local differences on frequencies above 430MHz.

**EFFECTIVE FROM 1st JANUARY 2015 UNLESS OTHERWISE SHOWN**

| 136kHz | NECESSARY BANDWIDTH | UK USAGE |
|---|---|---|
| 135.7-137.8kHz | 200Hz | CW, QRSS and Narrowband Digital Modes |

**Licence Notes:** Amateur Service - Secondary User. 1 watt (0dBW) ERP.
**R.R. 5.67B.** The use of the band 135.7-137.8kHz in Algeria, Egypt, Iran (Islamic Republic of), Iraq, Lebanon, Syrian Arab Republic Sudan, South Sudan and Tunisia is limited to fixed and maritime mobile services. The amateur service shall not be used in the above-mentioned countries in the band 135.7-137.8kHz, and this should be taken into account by the countries authorising such use. (WRC-12).

| 472kHz (600m) | NECESSARY BANDWIDTH | UK USAGE |
|---|---|---|
| 472-479kHz | 500Hz | CW, QRSS and Narrowband Digital Modes |

**Note 1:** It should be emphasised that this band is available on a non-interference basis to existing services. UK amateurs should be aware that some overseas stations may be restricted in terms of transmit frequency in order to avoid interference to nearby radio navigation service Non-Directional Beacons.
**Licence Notes:** Amateur Service Secondary User. Full Licensees only, with NoV. Note that conditions on power are specified by the NoV terms.
**R.R. 5.80B.** The use of the frequency band 472-479kHz in Algeria, Saudi Arabia, Azerbaijan, Bahrain, Belarus, China, Comoros, Djibouti, Egypt, United Arab Emirates, the Russian Federation, Iraq, Jordan, Kazakhstan, Kuwait, Lebanon, Libya, Mauritania, Oman, Uzbekistan, Qatar, Syrian Arab Republic, Kyrgyzstan, Somalia, Sudan, Tunisia and Yemen is limited to the maritime mobile and aeronautical radionavigation services. The amateur service shall not be used in the above-mentioned countries in this frequency band, and this should be taken into account by the countries authorising such use. (WRC 12).

| 1.8MHz (160m) | NECESSARY BANDWIDTH | UK USAGE |
|---|---|---|

| | | | |
|---|---|---|---|
| 5,362-5,374.5 | 12.5kHz | 5,362-5,370kHz – Digital Mode Activity in the UK | |
| 5,378-5,382 | 4kHz | | |
| 5,395-5,401.5 | 6.5kHz | | |
| 5,403.5-5,406.5 | 3kHz | 5,403.5kHz – USB Common International Frequency | |

Unless indicated, usage is all-modes (necessary bandwidth to be within channel limits).
**Note 1:** Upper Sideband is recommended for SSB activity.
**Note 2:** Activity should avoid interference to the experimental beacons on 5290kHz.
**Note 3:** Amplitude Modulation is permitted with a maximum bandwidth of 6kHz, on frequencies with at least 6kHz available width.
**Licence Notes:** Full Licensees only, with NoV. Note that conditions on transmission bandwidth, power and antennas are specified by the NoV terms.
Notes to the Band Plan. As on page 42.

| 7MHz (40m) | NECESSARY BANDWIDTH | UK USAGE |
|---|---|---|
| 7,000-7,040kHz | 200Hz | Telegraphy – 7,030kHz QRP (low power) Centre of Activity |
| 7,040-7,047 | 500Hz | Narrowband Modes (Note 2) |
| 7,047-7,050 | 500Hz | Narrowband Modes, Automatically Controlled Data Stations (unattended) |
| 7,050-7,053 | 2.7kHz | All Modes, Automatically Controlled Data Stations (unattended), (Note 1) |
| 7,053-7,060 | 2.7kHz | All Modes, Digimodes |
| 7,060-7,100 | 2.7kHz | All Modes, SSB Contest Preferred Segment Digital Voice 7,070kHz; SSB QRP Centre of Activity 7,090kHz |
| 7,100-7,130 | 2.7kHz | All Modes, 7,110kHz – Region 1 Emergency Centre of Activity |
| 7,130-7,200 | 2.7kHz | All Modes, SSB Contest Preferred Segment; 7,165kHz – Image Centre of Activity |
| 7,175-7,200 | 2.7kHz | All Modes, Priority For Inter-Continental Operation |

**Note 1:** Lowest LSB carrier frequency (dial setting) should be 7,053kHz.
**Note 2:** PSK31 activity starts from 7,040kHz. The narrowband modes segment starts at 7,040kHz.
**Licence Notes:** 7,000-7,100kHz Amateur and Amateur Satellite Service – Primary User. 7,100-7,200kHz Amateur Service – Primary User.
Notes to the Band Plan: As on page 42.

*Fig 6.3: Excerpt from the RSGB Band Plan, as published in RadCom.*

periods of low solar activity, 20m can be very quiet, with only semi-local stations around Europe to be found, though it is never too long on 20m before some DX signals return.

Long-path propagation is quite reliable on 20m when, instead of signals coming from the expected direction, they come in at 180° to that, the long way around the world. Look for signals from eastern Australia (especially VK4, VK2, VK1, VK3 and VK7), New Zealand (ZL) and, if you are lucky, other parts of the Western Pacific, typically for an hour or two after sunrise, coming in from the direction of South America.

## 7MHz – 40-METRE BAND

We skip over 10MHz, the 30m band, because by IARU recommendation there is no SSB activity on that band, to look instead at 40 metres. In most of the world the band is now 200kHz wide, from 7000 to 7200kHz. In the UK, the SSB part of the band is considered to be 7050 – 7200kHz although in the USA 7000 – 7125kHz is used for Morse and data mode transmissions and *not* for SSB. The RSGB Band Plan lists 7175 – 7200kHz as "Priority for Inter-Continental Operation" but DX stations generally operate anywhere between 7100 and 7200kHz.

In ITU Region 2 (North and South America) and in some parts of the Pacific 7000 – 7300kHz is allocated to amateurs, although in the rest of the world high-power AM broadcast stations operate between 7200 and 7450kHz.

During daylight hours, 40m supports only relatively short-distance working (from the UK, around the British Isles and into north and western Europe). Unlike all the higher-frequency bands, 40m is a night-time DX band: DX can be worked from an hour or two before sunset to the east, throughout the night, and for up to an hour or two after sunrise towards the west.

The grey-line, the twilight zone that separates the day and night sides of earth, can provide some excellent DX. See **Fig 6.4**. In this example

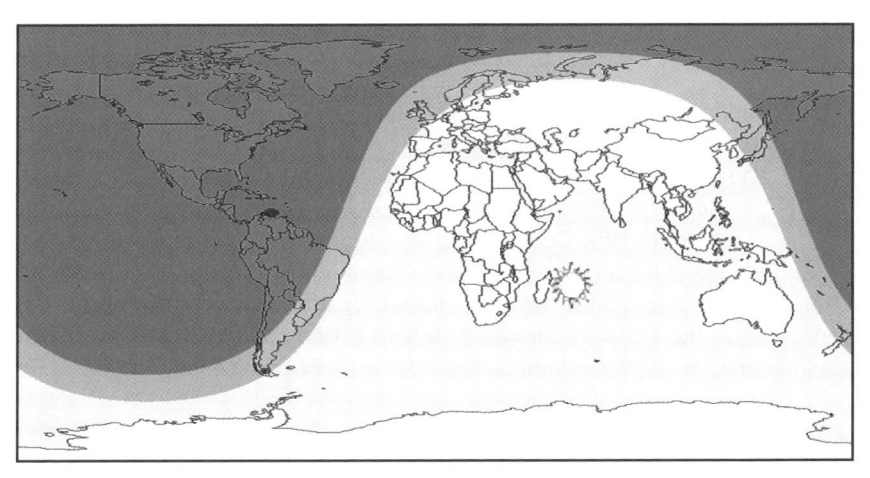

*Fig 6.4: Grey-line map as at 0800UTC on 21 December. The map clearly shows the daylight and darkness parts of the world and also the twilight zone or 'grey-line' along which DX may be worked.*

(0800UTC on 21 December), it is just after sunrise in the UK and the sun is overhead (i.e. it is midday) in the Indian Ocean. A 40m DXer in the UK might expect to be able to work DX stations towards the dark side of the earth, in this case westwards.

However, it can be seen from **Fig 6.4** that the whole of the British Isles is within the grey-line, which crosses the Atlantic and continues down the east part of Brazil, covering Uruguay, the eastern half of Argentina and the south of Chile, before crossing the Pacific Ocean. A 40m DXer might therefore reasonably expect enhanced signals from PY, CX, LU and possibly CE during this time. The grey-line continues across the North Island of New Zealand, the Solomon Islands and Vanuatu before crossing Japan, then going across northern Russia and Scandinavia before returning to the British Isles again. So not only can DX in South America be worked at this time, but also New Zealand, the western Pacific islands and Japan. The grey-line map also shows that countries such as Indonesia, the Philippines, Malaysia, Singapore and Thailand are in broad daylight (at 0800UTC on 21 December), so there is little to zero chance of making contacts with these countries even when Japanese and perhaps Korean stations are being worked.

## 3.5MHz – 80-METRE BAND

We skip over the 60-metre band (5MHz) frequencies, because they are not available to all licensees in the UK, and because there are many countries that have no allocation at 5MHz. We look instead at the 80m band, 3500 – 3800kHz, with SSB operation between 3600 and 3800kHz. However, other than in the major contests, most SSB DX activity is concentrated in the top 15kHz of the band, between 3785 and 3800kHz, the so-called 'DX Window'. In fact, the top 25kHz of the band, 3775 – 3800kHz, is considered to be "Priority for Inter-Continental Telephony (SSB) Operation" though many local and semi-local contacts can be heard taking place in that part of the band.

Unlike all the other HF bands, DX propagation on 80m does not peak with high solar activity and is often better at times of solar *minimum*, because there are less likely to be magnetic storms that cause severe attenuation of signals during periods of low solar activity.

During the day, 80m is only suitable for short-distance contacts (around the UK and perhaps into Belgium and France): 80m is very much a night-time DX band. Grey-line propagation (see **Fig 6.4**) is also important on 80m, with long-path New Zealand stations audible around sunrise in the UK when that coincides with sunset in New Zealand. Many true DX signals on 80m are very weak and the high levels of static from thunderstorms, particularly in the summer, makes 80m SSB DXing something of a specialist pursuit, with efficient low-angle antennas virtually a necessity.

## REFERENCES

[1] RSGB Band Plans: www.rsgb.org/main/operating/band-plans
[2] *Radio Auroras*, Charlie Newton, G2FKZ (revised edition, 2012), RSGB, available from www.rsgbshop.org

# 7 Working HF DX on SSB

IN THIS CHAPER WE WILL take a look at the techniques involved in working HF DX stations on SSB, but first examine how to find out when DX stations are likely to be on the air. Most of the monthly amateur radio magazines carry information about forthcoming DX activity, provided it has been notified to them in time for their publishing deadlines. The 'HF' column in *RadCom* and the 'How's DX?' column in *QST* are good examples.

The good old method of simply tuning the bands still exists of course but former printed DX newsletters have now been superseded by websites and e-mail bulletins. A good DX news website, which is well illustrated with photographs of the DX locations mentioned, is *DX-World* [1], run by Col McGowan, MM0NDX (see **Fig 7.1**). *QRZ DX* [2] and *The Weekly DX* [3] are weekly bulletins sent out by e-mail, but both are on subscription. *The Daily DX* [3] is, as its name suggests, published daily (on weekdays), again on subscription. All these bulletins contain information about forthcoming and

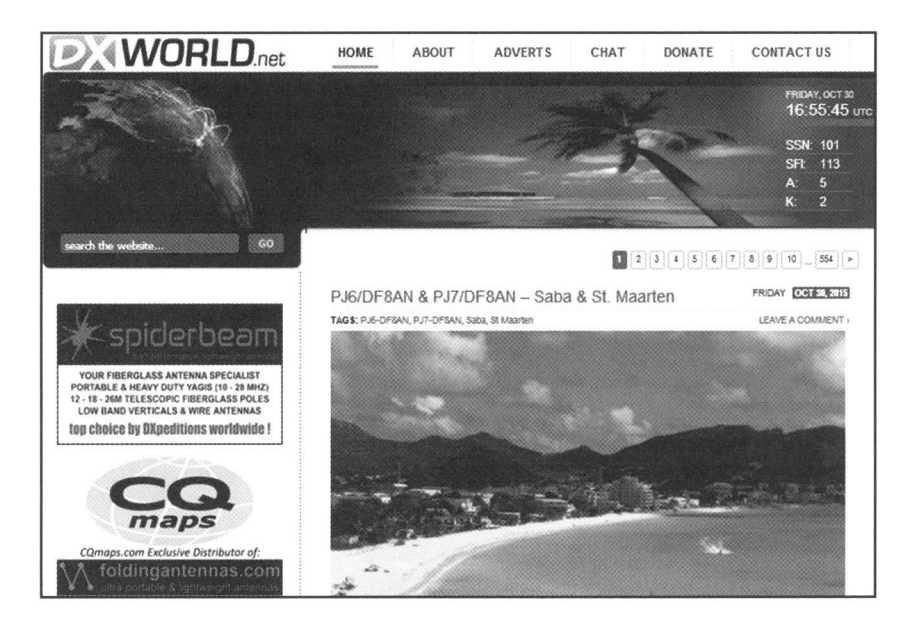

*Fig 7.1: The DX-World website often 'breaks' DX news.*

# HF SSB DX BASICS

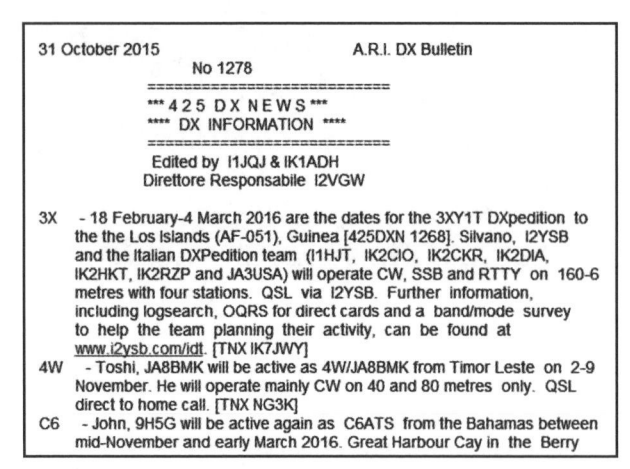

```
31 October 2015              A.R.I. DX Bulletin
                No 1278
==================================
            *** 4 2 5  D X  N E W S ***
            **** DX INFORMATION ****
==================================
         Edited by I1JQJ & IK1ADH
         Direttore Responsabile I2VGW

3X  - 18 February-4 March 2016 are the dates for the 3XY1T DXpedition to
     the the Los Islands (AF-051), Guinea [425DXN 1268]. Silvano, I2YSB
     and the Italian DXpedition team (I1HJT, IK2CIO, IK2CKR, IK2DIA,
     IK2HKT, IK2RZP and JA3USA) will operate CW, SSB and RTTY on 160-6
     metres with four stations. QSL via I2YSB. Further information,
     including logsearch, OQRS for direct cards and a band/mode survey
     to help the team planning their activity, can be found at
     www.i2ysb.com/idt. [TNX IK7JWY]
4W   - Toshi, JA8BMK will be active as 4W/JA8BMK from Timor Leste on 2-9
     November. He will operate mainly CW on 40 and 80 metres only. QSL
     direct to home call. [TNX NG3K]
C6   - John, 9H5G will be active again as C6ATS from the Bahamas between
     mid-November and early March 2016. Great Harbour Cay in the Berry
```

*Fig 7.2: Excerpt from a 425 DX News bulletin distributed by e-mail.*

current DX activity, after-the-event information on many operations, QSL information and lots of other information of interest to DXers. Finally, *425 DX News* **[4]** (**Fig 7.2**) is another weekly newsletter but free of charge to anyone who signs up for it.

Real-time DX information is available from the global Cluster network linked by the Internet. The Clusters carry real-time 'spots' of DX activity, solar data from WWV etc. A popular web-based Cluster is *My DX Summit* **[5]** run by the OH8X group in Finland. This collects Cluster spots from around the world and displays them in various ways according to which filters are selected. In **Fig 7.3** the 'HF' and 'Phone' filters have been selected, so spots made of stations using other modes, or on VHF, are not shown. You can also post your own spots and announcements on to the Cluster network via *My DX Summit*.

## FINDING THE DX

You may know from *DX-World* or a DX news bulletin that a DXpedition is taking place, but how to find it? The first aspect is to think about propagation. Which band or bands are likely to support propagation to that part of the world, and at what times? A great web-based resource is VOACAP DX

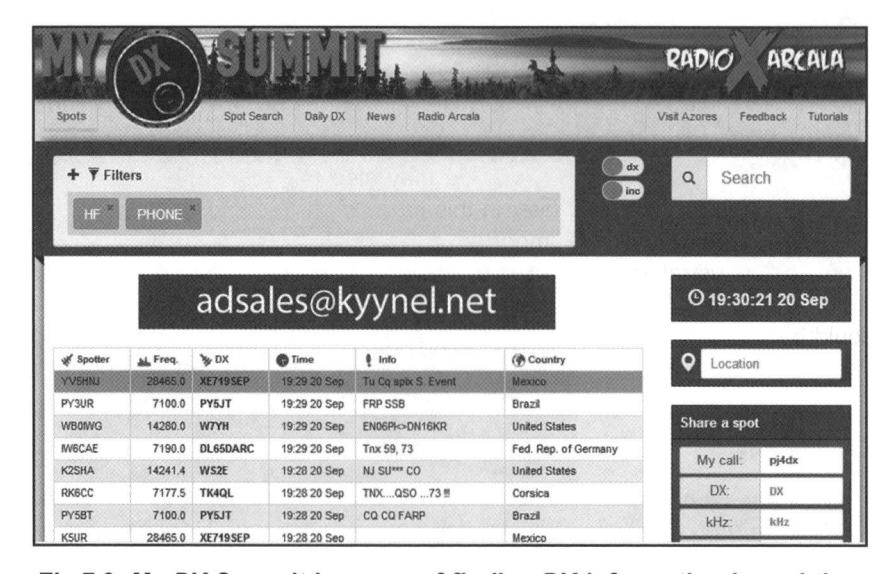

*Fig 7.3: My DX Summit is a way of finding DX information in real time.*

## 7 – WORKING HF DX ON SSB

Charts [6] which lists most of the pre-announced DXpeditions for the next six months or so. By inputting your own grid reference and simply clicking the "Run!" button, short-path and long-path propagation predictions are calculated for your own location to all the DXpeditions – see **Fig 7.4**.

The NCDXF beacon chain [7] can give a good indication of how propagation *actually* is, compared with what was predicted.

When you've worked out that propagation is likely to be favourable, it is a good idea to check what that means in terms of local time at the distant end. If it's a DXpedition you are chasing, they will probably be around whenever propagation is likely. But if we are talking about a local amateur, remember that on weekdays they will probably be at work and at night (their time) they will probably be asleep! Try to put yourself in their shoes.

DXpeditions tend to stick close to certain well-known frequencies. On SSB, 3795, 14185 – 14200, 18145, 21295, 24945 and 28495 are perhaps the most popular DXpedition frequencies although many expeditions will assume that nowadays most DXers have Cluster access and will therefore find them *wherever* they choose to operate in the bands. As mentioned in Chapter 6, IOTA DXpeditions can often be found around 14260, 21260kHz and 28460kHz.

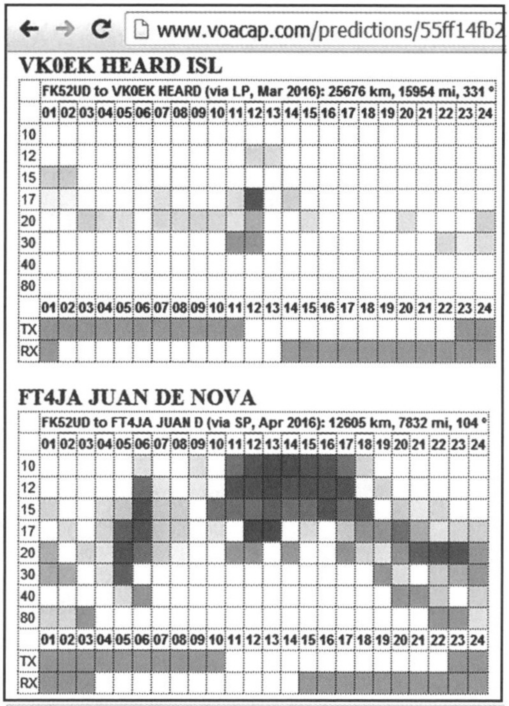

*Fig 7.4: VOACAP propagation predictions worked out from your location to various DXpedition locations. Available at the click of a button.*

From the UK, Pacific stations are often identifiable by having some 'flutter' on their signals, resulting from signals passing through the auroral zone. A local amateur, trying to avoid major pile-ups, might make a point of operating away from the more popular areas of the band, so may, for example, operate below 14150kHz on SSB. He may also take refuge in one of the popular DX nets (dealt with later in this chapter).

Operators' accents are important in alerting to you to a possible DX station. An Australian or Japanese accent is usually distinctive, while an American accent heard when the band is closed to the continental USA could be from somewhere like Hawaii or perhaps Guam.

Often an immediate pointer to a rare station is the existence of a pile-up. This is bad news in a way, as it means you are going to be in a competitive situation. It is far better to be the *first* person to find a DX station, when he calls CQ!

There is an alternative, which is to call "CQ DX" yourself, but this tends not to be terribly productive unless you have a big signal. However, some DX stations will prefer to call others, rather than call CQ themselves, so as to avoid getting into a pile-up situation, so it is always worth a try.

# HF SSB DX BASICS

## WORKING THE DX

So you have found the DX station on what appears to be a clear frequency, working stations that you can't hear. The temptation is to call him as soon as he ends a contact. Wrong, on several counts! The first thing is to determine what is going on. Spend a minute or two listening. It will pay dividends.

The chances are that the reason you can't hear the stations he is working is because he is operating split-frequency, in other words listening on a frequency other than the one he is transmitting on. Assuming he is a competent operator, he will announce his listening frequency often. If he is operating split, this will dictate how and where you call him (see the section later, on split frequency operation).

Keep any call short: normally give your callsign just *once*, using recognised phonetics. If he is working quickly through a pile-up, don't give his callsign: he knows it already! By giving your callsign just once, you are helping everybody. If you are the first station the DX station hears, he will come back to you immediately and give you a report. For example:

Him: "Two Echo Zero Alpha Alpha Alpha, 59".
You: "Thanks, also 59"
Him: "QSL, Echo Six Zulu Zulu QRZ?"

. . . and away he goes with the next contact. If 2E0AAA had called twice his second call would just have slowed things down. Or if E6ZZ (on Niue in the Pacific) had responded to someone *else's* first call, 2E0AAA would just have been causing unnecessary interference. If E6ZZ had only heard *part* of 2E0AAA's callsign, no problem, the gaps can be filled on the next transmission:

Him: "Two Echo Zero Alpha, 59"
You: "Roger, Two Echo Zero Alpha Alpha Alpha, 2E0 Alpha Alpha Alpha, also 59"
Him: "QSL, Two Echo Zero Alpha Alpha Alpha. Echo Six Zulu Zulu QRZ?"

In this example 2E0AAA *did* give his callsign twice, because it was clear that the DX station was calling him and also clear that the DX station was having problems picking out full callsigns. Also, if other stations were still calling, by giving their callsigns more than once, it is likely that 2E0AAA would have faced interference if he had only given *his* callsign once and so would have had to repeat again.

Some amateurs call with only a part of their callsign, typically the last two letters, such as "Alpha Alpha". This is bad operating and seems to have originated because DX stations often *respond* with a partial call. That's not actually because they *want* partial calls, it's simply that that is all they have been able to hear through the pile-up. By coming back with at least a *partial* call, rather than calling "QRZ?" again, they can press on with the contact rather than wasting time, assuming there was only one station in the pile-up with "Alpha Alpha" in the call. Assuming, also, that everyone else takes the hint and stands by which, sadly, isn't always the case. But if you were only to send a partial callsign rather than your full call (quite apart from the licensing issues, whereby you are required to identify yourself when making "calls to establish contact with another amateur") you are wasting time. If you only send a partial call and the DX station only hears part of that, he may end up with just a single letter, which is of no use at all. Even if he hears you clearly, it will still require you to give your full callsign on the next transmission.

Don't slow things down with unnecessary requests or questions, unless things are quiet and the DX station is repeatedly calling CQ with no response. If you need his QSL information, check on the Internet, for example on the QRZ.com site **[8]**, or wait for him to announce it. If you want to know when he will be on another band, wait for him to announce that, or leave him on the loudspeaker while you do something else in the shack, and wait for him to close on that band and move to another.

If you are used to exchanging pleasantries with other amateurs, these brief exchanges with DX stations can seem rather impersonal, but DXing is immensely popular and most DX stations and DXpeditions want to give as many people as possible the chance for a contact.

Perhaps the main piece of advice is always to take your cue from the DX station.

## WORKING BY NUMBERS & OTHER TECHNIQUES

If the pile-up gets very large, some DX operators may start 'working by numbers'. For example, he might call for "Only stations with Number One in the callsign". If he starts with One and you are an M0, it can be pretty frustrating to sit while he works his way through all the other numbers, but if callers step out of line all it does is slow things down and, in the extreme, the DX operator will give up and switch off.

I do not recommend the working by numbers technique, for several reasons. Firstly, as suggested, it *does* cause frustration among amateurs with the 'wrong' number in their callsign. It is particularly frustrating if, when you first tune in to him, he has just passed 'your' number. Secondly, it is unfair because some amateurs with two digits in their callsign get two bites of the cherry – a 2E0 could legitimately call when the DX station is asking for Number Zero *and* Number Two, whereas an M6 station does not have that luxury. Finally, and most importantly, propagation windows – especially to very distant stations – are often short and while you may have good propagation and be able to work the DX when you first hear him, if you have to wait 20 or 30 minutes before you can even make your first call, conditions could well have changed for the worse and you may have no chance of making the contact. Nevertheless, if the DX station chooses to use this technique there is nothing you can do about it other than to wait your turn. Calling out of turn only causes even more frustration by other stations and potentially leads to an ill-disciplined free-for-all. If you call out of turn the DX station will probably not log you anyway.

If you only have a modest station you may find you're getting nowhere in calling the DX – despite doing all the right things. If that's the case you

# HF SSB DX BASICS

could try to be first on frequency the following day by looking for the DX station around the same frequency, but earlier. A look at the spots reported on *My DX Summit* **[5]** will give a good indication of when the DX station is likely to be active.

If it is a big expedition, on the air for a week or two, you might just have to sit things out for a few days. As the number of callers diminishes towards the end of the operation there is always far less competition. Often DXpeditions, even to rare locations, have to call CQ and are said to be "begging" for callers towards the end of an operation.

But it is always worth hanging in there, as propagation may suddenly change and favour you, or the DX station himself, recognising that southern European stations (Italy, Spain, Portugal etc) have a clear propagation advantage over northern Europe, might switch the odds to your favour by announcing: "Please stand by everybody, UK only now" (or whatever).

However, decades of experience has shown that the best technique for handling pile-ups is split frequency operation, and so that deserves a section all to itself.

## SPLIT FREQUENCY

If you hear a DX operator making plenty of contacts, but you can't hear the callers, the chances are that you are listening to a split-frequency operation. Before calling *any* DX station on his own frequency, it's always worth listening for a while to determine whether he is working split: if you call him co-channel, you risk the wrath of others who have been waiting and who will almost certainly inform you of the error of your ways!

Most DXpeditions and many DX operators choose to operate split frequency. The concept is simple: by transmitting on one frequency and listening on another, callers can hear the DX station clearly, rather than through a mass of other callers, and therefore know when to go ahead with their contact and, more importantly, when to remain silent while another contact is taking place.

The DX station may choose to listen on a single spot frequency, or perhaps over a narrow band of frequencies in order to make it easier for the DX station to pick out an individual caller. If he chooses to listen on a single discrete frequency it is normally 5kHz higher than his transmitting frequency, although it could be as close as 3kHz. It should not be any *closer* than 3kHz, because stations calling less than 3kHz away would cause interference to the DX station and prevent others from hearing the DX well, which would defeat one of the purposes of the split frequency operation.

However, it could be that there are already other stations operating between 3kHz and 8kHz higher than the DX station's transmit frequency and if that is the case the DX obviously cannot listen 5kHz above his transmit frequency, because the stations calling would cause interference to the already-existing QSO. Instead, the DX station might nominate a frequency perhaps 10kHz higher than his transmit frequency if that is clear, or maybe 5kHz *lower* in frequency. As always, listen to the instructions from the DX station and follow his lead.

# 7 – WORKING HF DX ON SSB

If there are very many stations calling the DX station, instead of listening on a single discrete frequency, he might choose to listen over a narrow band of frequencies, say 5 to 10kHz higher or, with a *very* large pile-up, 5 to 15kHz higher. Spreading out the pile-up over 5 or 10kHz makes it easier for the DX station to pick out individual callers. Some operators, particularly those who are relatively inexperienced at DXpedition operating, find it very difficult to pick out an individual callsign, so instead of spreading the pile-up over 5 or 10kHz, they listen over a much wider range of frequencies. This is poor operating practice as this certainly *will* cause interference to other band users, causing resentment and in some cases even deliberate interference to the DX station (though this is never to be condoned).

Another poor operating technique on SSB is for the DXpedition operator simply to announce they are listening "up", but without specifying *where* up. This seems to be prevalent among those who are primarily CW operators but who are making an occasional excursion on to SSB. On CW it is normal to operate split by listening just 1kHz up, but by sending "up" (as opposed to "up 1") on CW the DX station expects the pile-up to spread out over, say, 1 to 3kHz, or even up to 5kHz higher. This technique does not work on SSB, though, as the pile-up is likely to be already calling 5 to 10kHz higher. By just saying "up" on SSB, in an attempt to call on a clearer frequency, stations will start calling first 11 or 12kHz higher, then if that fails, 14 or 15kHz higher and before long the DX station has an unnecessarily wide pile-up causing interference to other band users.

The DXpedition station should *always* announce their QSX (listening frequencies) if they are operating split, preferably after *every* QSO but certainly after every two or three QSOs. Unfortunately, not all do so as frequently as would be ideal. Failure to announce QSX causes stations to call the DX station on his own frequency, defeating one of the main objects of operating split in the first place. Suppose a DX station is transmitting on

***Ant, MW0JZE, operating split, 4W6A (Timor-Leste), September 2011.***

# HF SSB DX BASICS

14195kHz and operating split. Typically he might say "listening on 14200" or "listening 5 to 10kHz up". Sometimes he might just say "five to ten", meaning 5 to 10kHz higher in frequency, or "two hundred to two-oh-five" meaning 14200 to 14205kHz.

A few moments listening should allow you to find the callers a DX station is working. Listen a little longer and you may also determine a listening pattern: does the DX station always respond to callers on the same frequency or does he listen a little higher up the band after each contact, finally dropping back down the band and starting the whole process anew? Many DX operators only respond to callers sending their complete callsign so ensure you always send the complete call and not just the last two letters.

Unless the DX station is specifically taking 'tail enders', i.e. stations who call as the previous contact is coming to an end, never call over the top of a contact in progress but wait until the DX station signs and calls "QRZ?" or similar. Listening pays dividends, compared with simply calling at random. This is how experienced operators running low power can often get through more quickly than less experienced operators running high power.

Some DX stations deliberately try to favour more attentive callers, for example by suddenly announcing a spot listening frequency elsewhere in the band. If you're on the ball you can move there, call and work him, while others are still calling on the original frequency. DXpedition operators from continental Europe sometimes make a brief announcement in their own language, which is often an invitation for their compatriots to call on a specific frequency. If you can understand enough of that language to work out what frequency the DX station is listening on, you may find yourself only competing with, say, Dutch or Norwegian stations and not the whole of Europe!

## *HOW* TO OPERATE SPLIT

If a DX station *is* working split, what should you do? If your transceiver only has a single receiver you would need to utilise the transceiver's two VFOs, one to receive (on the DX station's frequency – let's call it VFO A) and the other to transmit (on what you hope will be the DX station's receive frequency – VFO B). Just about every transceiver available today has twin VFOs, so this technique can be used with any transceiver.

The difficulty is that when the DX station is listening not on one discrete frequency, but over a range of frequencies, you will not know on which frequency you should be transmitting in order for the DX station to hear you. You can of course listen on VFO B to hear who is calling the DX station, but when you are listening to VFO B you cannot hear what is happening on VFO A, i.e. the DX station himself.

What is required is for A and B not to be VFOs, but rather completely separate and independent receivers. Using stereo headphones, you can then listen either to RX A or RX B, *or* you can listen to RX A (the DX station) in your left ear, and RX B (the pile-up) in your right ear. When you hear the DX respond to a particular station, if you can use the second receiver to find that station, you know that that is the frequency the DX station is listening on. When the QSO is completed (and only then, not before!) you can call with the knowledge that, at that moment at least, the DX station is listening on that frequency – see **Fig 7.5**. This technique gives those operators with dual receivers a distinct advantage over those who only have twin VFOs. Unfortunately, very many operators now have transceivers with dual receivers and the advantage is not as great as it once was, as the majority of stations calling the DX may well be using the same technique!

The second receiver can still help, though. If you can hear that almost everyone is calling on the same frequency, the chances are the DX station will find it difficult to pick out a callsign from the *mêlée*. Using your second receiver, you might be able to find a frequency that is relatively clear of others calling, though still within the range of receive frequencies announced by the DX station. If you call on this frequency, it is likely that the DX station will hear you more clearly than if you were to try to battle it out with everyone else.

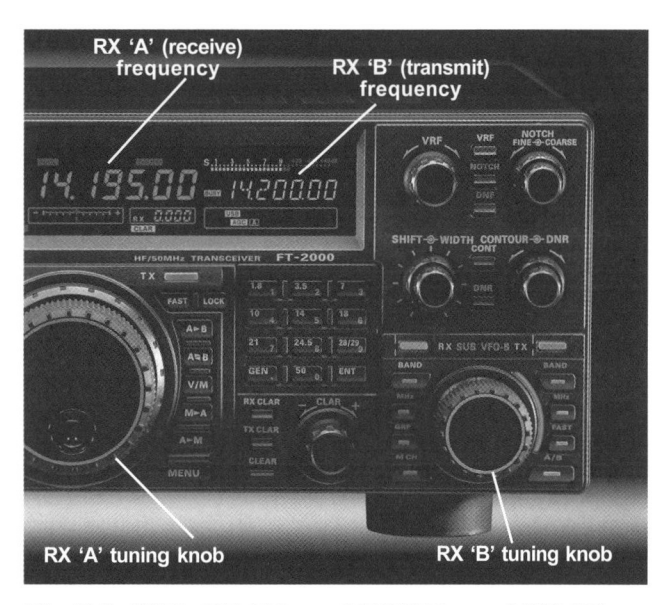

*Fig 7.5: With RX 'A' on 14195kHz and RX 'B' on 14200kHz, this transceiver is ready to go on split frequency operation.*

As mentioned briefly in Chapter 4, an alternative to twin receivers is a facility known as 'dual watch'. Here the audio from the transceiver's two receivers is combined, with separate AF gains or a balance control adjusting the relative volume of audio from each frequency, although there is no facility to hear the two receivers simultaneously but separately. There is a danger in a DXpedition station using dual watch to monitor his own frequency when working split, though. Because the audio from the two receivers is combined, the DX station cannot tell which of the two frequencies a station is calling on. If the DX should respond to someone who has called on his *transmit* frequency, it will soon lead to chaos as other callers will take this as their cue also to call on that frequency.

## NETS AND LIST OPERATIONS

Many nets are associated with a particular organisation such as the Royal Signals Amateur Radio Society, but in this section we look at nets and lists specifically run for the purpose of enabling participants to work DX.

It should be noted that some DXers consider nets and lists to be a form of cheating, in that you are enlisting the help of a third party (the Master of Ceremonies, or MC) to run the operation and keep other stations at bay while you make your call to the DX station.

DX nets meet on (or near: a DX net has no more right to the use of a particular frequency than any other amateur) frequencies that are usually well-publicised, and at regular times. Usually, at the beginning of the net, the MC will ask for DX check-ins, then other check-ins will be called. The MC will then go round each of the participants in turn, asking them if they wish to call any of the DX stations. When your turn comes, make the call and complete the contact as efficiently as possible, and hand back to the MC. There may well be a lot of sitting around waiting your turn, but when things work well it means you don't have to compete with other amateurs (potentially running higher power) all calling at the same time.

The participants in the DX net will know the callsign of the DX stations who have checked in (although this is no different than being alerted to the presence of a DX station by monitoring the Cluster), but under no circumstances should the MC or any other participant 'help' a contact to take place by giving the full callsign of the participant to the DX station, or by relaying a signal report in either direction.

Lists are similar to nets, but tend to happen spontaneously. Let's suppose a station appears from a rare entity, but using low power and a simple antenna. The demand for the station is such that, whenever he goes on the air, he is swamped by callers. He tries the obvious solution and operates split frequency (see previous section), but his signal isn't strong enough to compete with those causing interference on his own frequency and his rate of making contacts falls almost to zero. What he may do is to ask one of the stronger, better-equipped, stations to "make a list". He will take maybe 10 callers at a time, then indicate when each should call.

One problem with list operations is that, due to propagation characteristics, the MC will pick up stations that have a strong signal with *him* but which may have marginal or no propagation to the DX station they are trying to contact. At the same time, there may be many amateurs in a different part of the world that *can* copy the DX station well, but which are too weak with the MC to get on the list. Thus a lot of time is wasted by stations giving RS 33 reports to the DX station and struggling to receive the sent signal report while many stations who can copy the DX station well are not given an opportunity to make a contact at all.

## QRP DXING

Having a modest station should not preclude anyone from chasing DX and QRP DXing has been a growth activity recently. Organisations like the GQRP Club **[10]** have published many designs for home-built QRP transmitters and transceivers but until fairly recently there have been few commercial designs developed specifically for QRP operation (though most can have their power reduced to QRP levels if required). However, the Yaesu FT-817 and the Elecraft KX3, in particular, have recently become very popular with amateurs who like to try to work DX at low power levels.

First though: what, precisely, is meant by the term 'QRP'? It is generally taken to mean an output power level of 5 watts or less, although many increase this limit to 10 watts PEP output for SSB transmissions. (Note, though, that the definition of QRP for the CQ World Wide DX Contests is that the "total output power must not exceed 5 watts" – even on SSB.)

## 7 – WORKING HF DX ON SSB

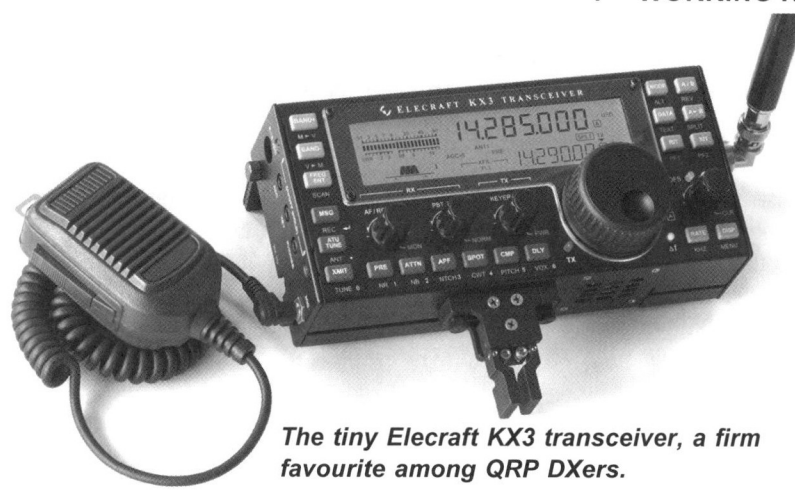

*The tiny Elecraft KX3 transceiver, a firm favourite among QRP DXers.*

QRP DXing is certainly harder than DXing with high power and requires more patience. It is perhaps more suited to experienced DXers who want a new challenge but those with limited-power licences should not be discouraged from chasing DX. It's not transmitter power, as such, that is relevant but the combination of that and antenna gain. So a UK or Australian Foundation licensee using 10W PEP output to an antenna with a 10dBi gain antenna (by no means impossible on HF when ground gain is also taken into account) will actually be *more* effective than a Full licensee running 100W output to a lossy trapped vertical or similar antenna.

There is plenty that can be done to maximise your chances of QRP DX success. For example, be active in the major DX contests, whether you consider yourself a contester or not. The bands will be full of activity, including plenty of DX stations, so the chasers won't all be after just one DX station. They will be scattered throughout the bands, divided between lots of DX stations to chase. So you shouldn't have so much competition, especially if you wait until well into the contest, for example the second day of a 48-hour event.

Also, the major DXpeditions set out to ensure that they work not only the best-equipped DXers, but also those with more modest stations. They recognise that many amateurs have restrictions on the sort of stations they can put together, but nevertheless want to participate in the DX game. The UK-led Five Star DXers Association expeditions, for example to East Kiribati (T32C, in 2011), are good examples, with round-the-clock operation on all bands, good monoband antennas, on a sea-front site, and on the air long enough to work their way through all the strongest callers and be left with more than enough time to ensure that QRP stations can also be worked.

Although QRP DXing may require more perseverance, remember that it is the QRO (high-power) station being called who is the one that has to have the receiving skills and equipment, as well as the patience, to pick out the weak QRP signal. Bear in mind that if a QRP station, running, say 10W PEP output, is only hearing a DX station running 1kW (legal in many countries although not the UK) at S8, then the high power station will only receive the QRP station at about S1, which could well be below his local

# HF SSB DX BASICS

## PITFALLS – WORKING DX
### Beware of the following pitfalls when chasing DX.

1. *Always* use phonetics when calling a DX station. Try alternatives (such as "Mexico" instead of 'Mike" if there is difficulty in understanding the standard phonetics, but *don't* use 'funny' phonetics.

2. *Never* call a DX station using only the last two letters of your callsign. *Always* send your complete call.

3. Only send your callsign *once* when making an initial call to a DX station. Should he *clearly* be responding to you, but has only received part of your callsign, you could then send your callsign twice if interference etc makes this necessary.

4. *Don't* repeat your callsign if the DX station already has it correctly. This only makes him doubt that he has copied it correctly the first time, especially if copy is marginal.

5. If the DX station responds to someone other than you, do *not* call again until he has completed that contact.

6. Beware of 'continuous calling': *only* call a DX station when he is listening for new callers (e.g. after saying."QRZ?")

7. If a DX station is operating split, *never* transmit on the DX frequency.

8. Only ask a DX station for information such as QSL instructions or when he will be active on another band *if* the DX station is *not* busy (e.g. calling CQ with no takers).

9. *Never* append /QRP ("stroke QRP") to your callsign: the DX station does not need to know your power level, only your callsign.

10. In a DX net or list operation *never* 'help' a participant by relaying information such as callsigns or signal reports.

11. Jargon such as "The personal is..." or (even worse) "The personal would be..." (instead of "My name is...") is considered poor operating. Use standard English wherever possible. The use of widely-recognised Q-codes such as QSL, QTH, QSO, QRX etc is of course acceptable and even encouraged in contacts with those whose first language is not English.

noise level (there is a 20dB difference between 10W and 1000W and, assuming 3dB per S-unit, S1 to S8 is a difference of 21dB). If a QRP operator is only receiving a high-power station at S6 or so, there is probably little likelihood of being heard at all. On the other hand, if the QRO station is being received at S9+20dB the QRP station might well be heard at S9.

A couple of final comments. Firstly, it isn't considered legitimate to call with high power to attract the attention of the DX station and then to drop to QRP to conduct the exchange of signal reports. Secondly, don't append "/QRP" to your callsign when calling stations. In the UK it is against licence conditions, as the only allowable suffixes are /A, /P, /M and /MM, but it is also counter-productive. Your time is better spent giving your callsign. If you want to chase DX as a QRP enthusiast, you have to be prepared to join in with everyone else and take your chances.

## QSLs AND QSLing

QSL cards have been around almost since the beginning of amateur radio itself and most DXers collect them, particularly if they actively participate in the DXCC or IOTA programmes where (with some exceptions), QSL cards are necessary to confirm that the contacts have taken place.

Most RSGB members will use the RSGB QSL Bureau **[10]** but, while

# 7 – WORKING HF DX ON SSB

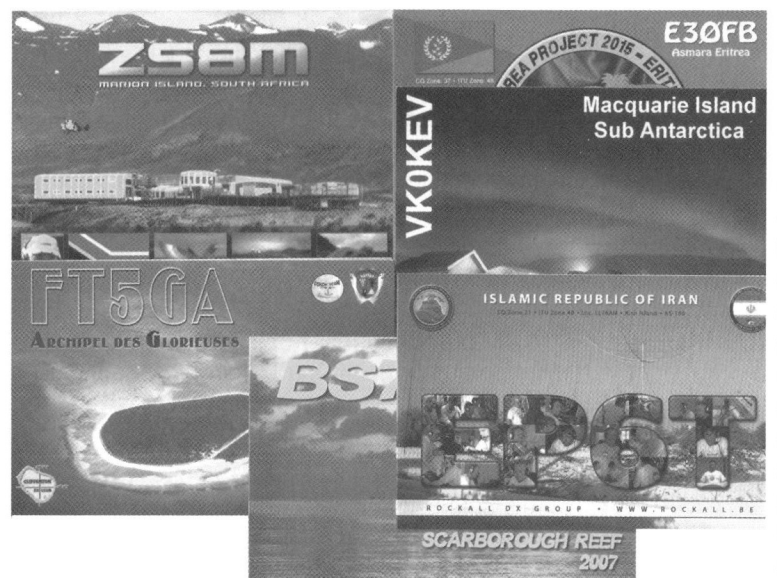

*The world's QSL bureaus are fine for receiving cards from those countries with large populations of amateurs, but less good when collecting cards from DX stations. For these countries you will usually need to request a card via a QSL manager or QSL direct.*

this is fine for collecting cards from countries with large populations of radio amateurs (Germany, Italy, USA, Russia, Japan etc), it is actually not a very efficient means of collecting cards from DX stations. Certainly use the QSL bureau in order to receive cards from those 'easy' countries, but do bear in mind that very many DXCC entities are in countries that simply do not have a functioning QSL bureau system. If you send a card to the RSGB QSL Bureau after you work any of the resident operators in Vietnam, Papua New Guinea, Egypt, St Kitts and Nevis, Azerbaijan or any one of very many other 'semi-rare' DXCC entities you are going to be disappointed, because those countries do not have a QSL bureau so I'm afraid you won't get a reply.

Even if the country concerned *does* have a QSL bureau, if the station you work is not a long-term resident in that country there is no point in sending the card through the bureau as he is unlikely to receive it. Some resident amateurs who are not members of the national amateur radio society also cannot receive cards via their QSL bureau.

The trick is *always* to find out the 'QSL information' of the station you work *before* sending off your card to the RSGB QSL bureau. Fortunately these days that is easy with the introduction of websites specifically for this purpose, the best known of which is QRZ.com **[8]**. A typical entry on QRZ.com will tell you (a) whether the station QSLs at all (some do not), (b) their postal and e-mail addresses, (c) whether they have a QSL manager, (d) whether they accept QSLs via the bureau or only direct, (e) if direct, whether they accept IRCs (International Reply Coupons, now being phased out in many countries including the UK) or US dollars (so-called 'Green Stamps'), and (f) whether they upload their logs to the ARRL Logbook of The World facility (see Chapter 2). Often there are also photographs and a lot more information besides.

If you are going to QSL direct you *must* enclose sufficient return postage for the DX station to reply. This is usually US $2 but in some countries this

91

# HF SSB DX BASICS

is insufficient to cover the cost of a stamp and you might need to send $3: QRZ.com will usually tell you. You *must* also enclose a self-addressed envelope for the DX station to return his QSL card to you. With the cost of airmail postage around £1.00 (depending on the country and the eventual size and weight of your letter), $2 or $3 for return postage, plus the cost of two envelopes, you can see that getting direct QSLs from 100 DXCC entities might cost around, say, £275, not to mention the cost of having your QSL cards printed in the first place.

Fortunately, there are alternatives! The first is *OQRS*, standing for Online QSL Request Service, which involves filling in the QSL request and QSO details on an online form. Most DXpeditions that upload their logs to *Club Log* [11] offer OQRS. Either bureau or direct cards can be requested by OQRS. Because DXpeditions rarely want their incoming QSL cards, requesting a bureau card by OQRS makes a lot of sense, firstly because you do not have to send your card, saving the cost of printing and postage to the QSL bureau and, secondly, halving the turn-around time because the DXpedition's QSL manager has your QSL request immediately, rather than having to wait months or even years for your QSL card to arrive through the bureau. OQRS bureau cards are invariably sent free of charge but you will still have to lodge stamped self-addressed envelopes at the RSGB QSL bureau to receive them and it can still take months – sometimes many months – for the card to arrive.

OQRS can also be used to send your QSL card by direct mail, usually at a cost of about US $3, to be paid by *PayPal*. Again, this saves you the cost of printing your card, posting it to the bureau and buying envelopes. At $3 per DXCC entity, 100 confirmations would cost about £200, a considerable saving but still a lot of money for many people.

It is for this reason that the ARRL's Logbook of The World [12], which is free of charge, is becoming ever more popular (see Chapter 2). LoTW will never replace traditional paper QSL cards entirely, because many DXers will still want to collect them as 'souvenirs' of their most memorable contacts, but for those mainly interested in increasing their DXCC score, LoTW is a boon and I see its use only increasing in the future.

## REFERENCES

[1] *DX-World*: www.dx-world.net
[2] *QRZ DX*: www.dxpub.com
[3] *The Weekly DX* and *The Daily DX*: www.dailydx.com
[4] *425 DX News*: www.425dxn.org
[5] *My DX Summit*: new.dxsummit.fi
[6] VOACAP DX Charts: www.voacap.com/dx.html
[7] NCDXF beacons: www.ncdxf.org/beacon/beaconschedule
[8] QRZ.com website: www.qrz.com
[9] GQRP Club: www.gqrp.com
[10] RSGB QSL Bureau: http://rsgb.org/main/operating/qsl-bureau
[11] Club Log: www.clublog.org
[12] LoTW on the ARRL website: www.arrl.org/logbook-of-the-world

# 8  Being DX

HAVING 'CHASED' AND WORKED a few DXpeditions, you might want to try your hand at organising your own, either as part of a family holiday (the so-called 'holiday expedition') or with a group of like-minded amateurs. An IOTA DXpedition, possibly to one of the islands off the coast of the British Isles, is ideal for starters but, with low-cost air fares and small, lightweight transceivers and even linear amplifiers and beams (see Chapter 3), it has never been easier to mount DXpeditions to overseas destinations. Many of these are in demand, even if they do not appear in the 'Most Wanted' surveys such as those described in Chapter 2. A lot of fun can be had by operating from DXCC entities such as Corsica, Madeira, Crete or Malta, not forgetting Jersey, Guernsey and the Isle of Man. The Channel Islands may be considered fairly commonplace within Europe, but they are among the rarest of the European DXCC entities in Japan, for example.

If you would like to give DXpeditioning a go yourself, take a look at the *World Licensing and Operating Directory* [1], which is full of advice on how to get licensed abroad and where to operate from when you get that licence. I compiled and wrote the book in 2008 and although it is now a few years old, much of the information in it remains equally relevant today.

## IOTA DXPEDITIONS

IOTA DXpeditions are an ideal way to get started in DXpeditioning. Many IOTA islands lie within a few hours' reach and can be put on the air relatively easily. **Table 8.1** lists all 28 IOTA Groups within the British Isles in their order of 'rarity'. Some of these – notably Rockall! – are remote and difficult to reach but others, such as Arran or Bute in the Scottish Coastal Islands Group (EU-123), or even the Isle of Wight in the English Coastal Islands (EU-120) are easy to get to. For a list of all the islands included in these groups, refer to the IOTA website [2] or the *IOTA Directory* [3].

Those amateurs who are able to activate a rare or even semi-rare IOTA Group can expect to generate big pile-ups and make thousands of contacts even during a short two or three day operation. Not *all* rare groups are remote and difficult to access: even in Europe and North America there are many that are needed by the chasers. A list of the most wanted IOTA Groups in each continent, ranked by rarity, is published in the *IOTA Directory* [3].

## BRITISH ISLES IOTA GROUPS

| # | IOTA | Name | % | # | IOTA | Name | % |
|---|---|---|---|---|---|---|---|
| 1 | EU-189 | Rockall | 9.8% | 15 | EU-099 | Les Minquiers Is | 54.1% |
| 2 | EU-112 | Shiant Is | 34.2% | 16 | EU-011 | Isles of Scilly | 59.3% |
| 3 | EU-111 | Monach Is | 34.6% | 17 | EU-123 | Scottish Coastal Is | 61.8% |
| 4 | EU-108 | Treshnish Is | 35.4% | 18 | EU-006 | Aran Is | 63.6% |
| 5 | EU-118 | Flannan Is | 37.2% | 19 | EU-010 | Outer Hebrides | 68.5% |
| 6 | EU-059 | St Kilda | 39.0% | 20 | EU-120 | English Coastal Is | 69.0% |
| 7 | EU-010 | St Tudwal's Is | 39.8% | 21 | EU-008 | Inner Hebrides | 70.1% |
| 8 | EU-109 | Farne Is | 40.7% | 22 | EU-012 | Shetland | 71.6% |
| 9 | EU-124 | Welsh Coastal Is | 46.8% | 23 | EU-009 | Orkney | 71.9% |
| 10 | EU-007 | Blasket Is | 47.2% | 24 | EU-013 | Jersey | 89.0% |
| 11 | EU-121 | Irish Coastal Is | 49.4% | 25 | EU-116 | Isle of Man | 91.2% |
| 11= | EU-103 | Saltee Is | 49.4% | 26 | EU-114 | Guernsey Group | 91.7% |
| 13 | EU-092 | Summer Is | 50.9% | 27 | EU-115 | Ireland | 92.9% |
| 14 | EU-122 | Northern Irish Coastal Is | 52.1% | 28 | EU-005 | Gt Britain | 93.5% |

*Table 8.1: All the IOTA groups of the British Isles including Ireland. They are ranked in order of 'rarity value': the percentage figure refers to the number of IOTA participants that have claimed that IOTA reference (as of September 2015).*

## DXPEDITION OPERATING

When you have many callers because you are the 'rare' station, it is up to to take control, or it may quickly get out of hand. There are several methods available: the first is to start operating split, as described in Chapter 7. This time, though, it is you who stays in one place and ask others to call you away from your transmit frequency. When it becomes clear that you will need to operate in this manner, ask the pile-up to stand by while you look for a suitable frequency to listen on. It should be close by, usually about 5kHz higher than the frequency you are operating on. But *do* find a clear frequency first, rather than moving your pile-up on top of an existing contact. Now the callers should be able to hear you clearly, so that your instructions to the pile-up should be understood and followed.

Try to pick a complete callsign out of the pile-up within the first few seconds or, if not a full one, enough of a call that there is no doubt who you are answering. If you can only copy a partial callsign, never work anyone other than the station you initially answered, e.g. if you go back to "Alpha Alpha", don't then work Papa Juliet Four Delta X-Ray! If you are heard to do this, other callers will take it as the green light for a free-for-all.

Do ensure you transmit regularly: if you stay silent for long periods the callers will engage in longer and longer calls, just adding to the level of interference.

If it is still too hectic, you can spread out the callers over a range of frequencies, though keep it to no more than 5 or 10kHz or you will start to take over the band. Even a 10kHz split is really only acceptable if you are on a major expedition to a rare spot.

Another technique is to make *directional* calls, by standing by for a particular continent or a part of a continent. For example, if you are operating from Europe (or close to Europe), the European stations will be workable for many hours at a time, whereas band openings to North America or Asia may be very short. So ask European stations to stand by while you check for calls from other continents.

If you are getting many callers from the East Coast of the USA it is worth taking a listen for the West Coast (or vice versa if you happen to be operating from the Far East or Pacific). Many US operators make a point of giving their State with their signal report, for example "Five Nine North Carolina", a useful habit since the numeral in US callsigns now often gives no indication of the location of the station.

With a less rare location, split-frequency operation may not be necessary, but you should still keep your operating crisp and avoid unnecessary exchanges of information. Give your callsign with every contact, or at least every two or three contacts, but there is usually no need to give out a lot of information such as name, equipment (the 'working conditions'), weather, QSL information etc. You can have longer chats when you are operating from home but, when you are sought after, it is only polite to try to accommodate as many of the callers as possible.

Try to avoid being drawn into nets or list operations if possible as they increase the overall number of exchanges that need to take place and hence reduce the number of contacts that can be made in the time available. One of the few times that it may be necessary is if you have limited battery power and therefore want to keep your transmit time to an absolute minimum. It may then be reasonable to ask for someone to maintain a list of callers for you. Ideally the list taker should be in the same part of the world as you are: It does not work when, for example, a European station takes a list of other Europeans on behalf of a weak Pacific station. He will hear those other Europeans who are strong with him on short-skip (quite possibly because they have high-angle antennas) but who may have zero chance of working into the Pacific. On the other hand, he may well not hear the well-equipped European station in his dead zone who could easily have made the contact.

> **PITFALLS – BEING DX**
> *Beware of the following pitfalls when operating as DX.*
>
> 1. *Don't* operate anonymously: announce your callsign after every QSO, or at least every two or three QSOs.
> 2. *Don't* 'work by numbers', it frustrates people with the 'wrong' number. Instead operate split if the pile-up is too much to handle co-channel.
> 3. *Always* announce *where* you are listening when operating split, *don't* just say "up":
> 4. When operating split keep your split narrow, preferably 5kHz and not more than 10kHz.
> 5. Stick to a rhythm: *don't* remain silent for too long as this encourages continuous calling.
> 6. *Do* stand by frequently for callers from other continents who might otherwise not be heard.

## REFERENCES

[1] *World Licensing and Operating Directory*, Steve Telenius-Lowe, 9M6DXX (PJ4DX), RSGB 2008, available from www.rsgbshop.org
[2] IOTA website: www.rsgbiota.org
[3] *IOTA Directory*, edited by Roger Balister, G3KMA, RSGB 2014, available from www.rsgbshop.org

# INDEX

*425 DX News*,   80
Adonis mics,   62
ALC,   62-63
Alinco DX-SR9,   55
antennas,   27-44
AOR ARD9800 'Fast Radio Modem',   19
ATU (internal automatic),   53-54
auroral propagation,   73-74
band plans,   72, 76-77
bandscope,   48
*Club Log*,   20-21, 24-25
*Daily DX, The*,   79
Deleted List (DXCC),   22
Digital Noise Reduction (DNR),   57
Digital Voice (DV),   19
digital voice keyer (DVK),   68-69
DSP,   14, 16, 57
dual watch,   58, 87
DX Cluster,   80
*DX Magazine, The*,   20
DX nets,   87-88, 95
*DX-World* (website),   79-80
DXCC List,   22
DXCC,   22-24
DXpedition operating,   94-95
Elecraft KX3,   50-51, 88-89
F2 propagation,   71-72, 74
filter method (of SSB),   14-15
FlexRadio 6700,   47
G(M)3SEK (Ian White),   65
G2FKZ (Charlie Newton),   74
G3SJX (Peter Hart),   47
G3TXQ (Steve Hunt),   43
G4GUO (Charles Brain),   19
G4JNT (Andy Talbot),   19
GQRP Club,   88
grey-line propagation,   77-78
Heil HC-4 & HC-5 mics,   61-62
Hertz,   6-7
Icom IC-703,   50-51
Icom IC-756Pro,   55
Icom IC-7700,   51
Icom IC-7851,   55
IF DSP, (see DSP)
International Telecommunication Union (ITU),   5
IOTA Foundation (IOTAF),   26
Islands On The Air (IOTA),   25-26, 81, 93-94
K9EID (Bob Heil),   61
Kenwood TS-480HX,   53

Kenwood TS-930S,   56
Kenwood TS-990S,   48
list operations,   87-88, 95
Logbook of The World (LoTW),   23-25, 92
long-path propagation,   77-78
Marconi, Guglielmo,   21, 34
MFJ-434B Digital Voice Keyer,   69
MFJ-653 Speech Articulator,   66
microphones,   60-62
MM0NDX (Col McGowan),   79
monitor facility,   69-70
MW0JZE (Ant David),   85
*My DX Summit*,   80, 84
NCDXF beacon chain,   81
nets,   87-88, 95
noise blanker,   56-57
NU9N (John Anning),   67
OQRS (online QSL request service),   92
phasing method (of SSB),   14-16
pitfalls (in DXing),   90, 95
QRP DXing,   88-90
*QRZ DX* (newsletter),   79
QRZ.com (website),   91
QSLs, QSLing,   90-92
receiver performance,   46-47
Shure mics,   62
Software Defined Radio (SDR),   47-48
solar cycles,   71-78
speech processing,   63-65
split operation,   58, 84-87, 94-95
Sporadic E propagation,   73-74
SSB (generation, reception),   14-19
Ten-Tec 715 RF speech processor,   65
TFT (thin film transistor display),   55
transmission audio tailoring,   65-66
transmission bandwidth,   67-68
twin receivers,   58
VOACAP DX Charts,   80-81
W1CBD (Clinton B DeSoto),   21-22
W2IHY Technology equipment,   66
WA6AUD (Hugh Cassidy),   20
waterfall display,   48
*Weekly DX, The*,   79
working 'by numbers',   83, 95
*World Licensing & Operating Directory*,   93
World Radio Conference (WRC),   5
Yaesu FT-450,   54
Yaesu FT-817,   50-51, 88
Yaesu FT-857D,   49, 65
Yaesu FT-2000,   49, 67